THE
NAVY
COMPANION

THE
NAVY
COMPANION

AN ILLUSTRATED COMPENDIUM OF MARITIME TERMS AND TRADITIONS

CEDRIC W. WINDAS

DOVER PUBLICATIONS, INC.
Mineola, New York

Bibliographical Note

This Dover edition, first published in 2019, is an unabridged republication of *Traditions of the Navy*, originally printed by Our Navy Inc. Publishers, Brooklyn, New York, in 1942.

Library of Congress Cataloging-in-Publication Data

Names: Windas, Cedric W., 1888–1966, author.
Title: The Navy companion : an illustrated compendium of maritime terms and
    traditions / Cedric W. Windas.
Description: Dover edition. | Mineola, New York : Dover Publications, Inc., [2019]
    | Originally published: Brooklyn, New York : Our Navy Inc., 1942.
Identifiers: LCCN 2019007727| ISBN 9780486836591 | ISBN 0486836592
Subjects: LCSH: United States. Navy—Sea life. | Naval art and science—
    Terminology. | English language—Terms and phrases.
Classification: LCC V736 .W5 2019 | DDC 359.1—dc23
LC record available at https://lccn.loc.gov/2019007727

Manufactured in the United States by LSC Communications
83659201   2019
www.doverpublications.com

TO MY WIFE

The best little navigator on
Life's uncharted seas that a
man ever sailed with.

*   *   *

Pillow my head on your breast, wife;
Soft arms hold me tight:
Ports that are half way 'round the world
Will be callin' me tonight.

                    C. W. W.

INTRODUCTION

MOST POWERFUL WEAPON of the United States Navy is the tradition of duty, sacrifice, and dauntless courage that has been handed down unbroken from John Paul Jones to the youngest apprentice seaman of today. This tradition has been preserved and enhanced by men whose names will live forever. Their spirit is the spirit that fires the Navy today.

The inheritance that the officers and men of the Fleet have received from the past is expressed in many forms. Particularly is it expressed in the ceremony, tradition, nautical customs, distinctive dress, insignia, and peculiarities of speech that set the Navy man apart from his fellow warriors.

A better knowledge of the origin of these traditions and customs is of real value not only to the Navy man but to anyone interested in the naval service. This book is a repository of such knowledge. Both the informative text and the excellent illustrations of Mr. Windas' book provide a wealth of authoritative information in a highly palatable form.

F. J. HORNE,
Vice-Admiral, U. S. Navy
Vice Chief of Naval Operations

CONTENTS

LIST of ILLUSTRATIONS

FOREWORD

TRADITIONS OF THE NAVY! What a multitude of Faith, Hope and High Endeavor expressed in those four words; what an avalanche of Romance.

Step up, gentlemen! and watch the Big Review of the Ships roll by: see the reckless Viking ships, the Greek triremes; mighty Spanish galleons and plucky little caravels . . . there goes a slab-sided '74, roaring around the Horn on her way into History.

Pipe the next in line! With canvas fore and aft, paddle wheels on either side, and smoke belching from a skinny funnel that I'll swear is thirty feet tall. Mark her well, for she is the connecting link between 5,000 years of Sail and little more than a century of Steam.

Here come the moderns! Slim destroyers, grim gray cruisers and arrogant battle-wagons. Floating fortresses of steel, half-smothered in spume and spray.

The navies of the world, gentlemen, all down through the Ages, each adding its quota to make up our Traditions.

*   *   *

Destroyer, cruiser, battle-wagon; what a flood of memories they unleash.

Crowded ports and swaggering sailormen; spice-laden breezes off Java Head . . . the smell of the pines in Christiania fjord . . . bronzed gods riding the surf at Waikiki . . . remember the shore-side fuss the day the old "Panay" got hers? I wonder if that girl in Melbourne is still waiting for me? I told her I'd be back . . . whatever became of "Slim" Smith? . . . Remember the night he came aboard with a half-share in Lotta's Fountain that a swab on the Embarcadero had just sold him for two dollars? . . .

* * * * *

But I'm way off my course: we were talking about Traditions of the Navy, their Customs and Terms, Superstitions and Nicknames; all handed down from one seaman to another like a jealously guarded heritage.

These same Terms, their origins ofttimes almost lost in antiquity, so enrich our common speech that rough sayings from rough tongues of other days, now find a snug harbor in our highly respectable standard dictionaries.

Their Customs, like the ships and the men, have changed but little; and the Sea . . . not at all.

I have gathered the explanations of these Traditions between the covers of this book, hoping that they will interest other seamen as they have interested me.

Above all is the hope that in the reading you will visualize the vast fleets and the gallant crews that drove them, braving the dark waters and challenging the Fates to find out what lay just beyond the Far Horizon.

Viking and Greek, Briton, American, French, Dutch and Norwegian; all members in good standing in the Ancient and Honorable Order of Deepwatermen, they have left us their Traditions.

Lord! keep their memory green.

C. W. W.

Hollywood, California

TRADITIONS of the NAVY

In the compilation of a volume such as this, one must be wholly alive to the fact that once in a while will come a challenge as to the authenticity of a certain item.

No one is quicker to protest "that's not the way I heard it" than the average seaman, and this in itself spells grief to any conscientious recorder.

Therefore, when in the course of research and preparation there have cropped up two explanations of one item, the writer has chosen what seemed to him the most logical and practical, for above all things a good sailorman is practical.

However, in the main, seaman and recorder alike concede that the explanations of the customs and terms contained in this book have a basis of fact which is undeniable.

Gleaned from Admiralty archives, from Navy records and from the tomes of famous historians ancient and modern, the contents of this volume were collected and illustrated for the entertainment and information of all lovers of the sea.

• CHIEF HOUSEMAID •

Old British Navy slang for the first lieutenant, because that officer was responsible for the cleanliness of the ship between decks.

• TARPAULIN MUSTER •

Old Navy slang for helping a shipmate in distress. A tarpaulin was rigged as a catch-net and the crew would file past, contributing whatever they could spare to help their financially embarrassed comrade.

• SKULL AND CROSSBONES •

Here's another boyish belief gone by the board; it appears now that the Skull and Crossbones was not the death-flag of the pirates. When this insignia was flown it meant that if you would deliver up all your cargo, you and your ship could go free. When total death and destruction was planned the corsairs would hoist the RED flag.

• THE JINX SHIP •

Some ships seem fated to sinister fortune, and a whole chapter of tragedy could be written about the old USS *Somers*. One hundred years ago Midshipman Spencer (son of the then Secretary of War) and two of his comrades were hanged at her yard arm for mutiny. A harsh sentence, for they were only boys. Of the men who decreed the death penalty, one died violently while horseback riding; one filled a drunkard's grave; one committed suicide; and one went insane. Bad luck continued to haunt both vessel and personnel, for her commander and twenty men were lost when she finally struck a reef and sank in deep water.

WINDAS

· WORKING FOR A DEAD HORSE ·

This expression, meaning that one is working at some job for which he has already been paid, dates back to early days, when it was customary to advance a sailor his first month's pay. After the ship had been thirty days at sea, the crew would construct a horse out of a barrel and odd ends of canvas. It would be hoist overside and set on fire, and as it drifted astern, the men would sing the old chantey "Poor Old Horse." This indicated that they would once again be working for wages, and not alone for "Salt Horse" (food).

· FIRST NAVAL COLLEGE ·

The first naval college of which there is any record was established at Sagres Portugal, in 1415 A.D. Here King Henry (known as Henry the Navigator) called together famous seamen from every nation to instruct his young countrymen.

· PENNSYLVANIA ·

Contrary to the general belief, the state of Pennsylvania was not named for its founder, William Penn, but for his brother Admiral Penn, as a mark of the King's favor.

· NELSON'S EYE-SIGHT ·

The famous statue of England's brilliant sea-lord which stands in Trafalgar Square London, is, oddly enough, not true to history. It depicts the admiral as being blind in his right eye, when as a matter of record, it was his left eye that was sightless.

• BLOW THE MAN DOWN •

Just a "Down-East" way of saying "*Knock* him down."

• KISSED BY MOTHER CAREY •

This is another way of saying, "Once a sailor, always a sailor." It is an old superstition that all boys who go to sea were kissed in their infancy by the sailorman's guardian angel Mata Cara (Mother Carey).

• BIG-WIGS •

This slang term for men in high positions gathers its meaning from the fact that senior officers in the old British Navy actually did wear huge wigs.

• GOING UP THE LEE LADDER •

It is an old tradition which demands that a seaman climbing a poop or bridge ladder shall ascend on the lee side.

This is because, in the old days, cleanliness of person was not considered so imperative as in modern times, and the officer of the watch objected to the smell of the sailors, of which he would receive a staggering reminder if they passed on his weather side. Today it is doubtful if any other body of men in the world are as clean as an average ship's company: but tradition dies hard, and because of it, seamen will probably pass to loo'ard of their officers, till Davey Jones pipes the Last Chantey.

· BLUE MONDAY ·

The term "Blue Monday" came into being as early as the 18th century. It originated because of an old custom aboard ships, whereby a man's misdeeds were logged daily, and the culprit flogged weekly, on the following Monday.

· DINGHY ·

This name for a ship's smallest boat, is a contribution to our nautical vocabulary from India. Dinghy means "small", and from this same word we get also our slang term "Dinky."

· SHIP'S MASTER ·

The official title for the sailing-master of a ship was created centuries ago, during the Punic Wars, when he was known as "Magestis Navis."

· HAWSER ·

The word "Hawser" is derived from the old English "Halter," meaning a rope for the neck.

• RANK OF ADMIRAL •

Strangely enough, there was no rank of Admiral in the U. S. Navy until July 1862.

• LEND A HAND •

LEND a hand, in sea-going parlance, is a request for help. BEAR a hand is a direct order.

• IDLERS •

This traditional name for members of the ship's company who stand no regular watch is a flagrant thrust at the toil-worn members of the medical staff, the steward, messboys, and others who work on a different schedule.

• AT LOGGER-HEADS •

This term, descriptive today of the angry relationship between two parties, dates back to the days when implements known as logger-heads were used in spreading hot pitch along deck-seams.

The tool consisted of a wooden pole with an iron head something like a flattened adze.

Men doing this work for long hours were apt to grow nerve-raw and quarrelsome, and their logger-heads made effective and ugly weapons.

A fight with these tools was a deadly knock-'em-down-and-drag-'em-out foray, and the seriousness of the affair is the general inference in the phrase "they are at logger-heads."

WINDAS

∘ CROSSING THE LINE ∘

This traditional ritual, now introducing the greenhorn to King Neptune in fun and merriment, was originally a very serious procedure amongst the Vikings, and was practised with all kinds of severe tests to see if the novice could really stand the hardships of the ocean.

PRESENT ∘ ARMS∘

The 'present arms' salute was originally a pacific and friendly gesture, meaning literally 'Presented for you to take if you wish'.

THE OLD "RED DUSTER"

The British flag was first named the 'Union Jack' in the reign of Queen Anne. Even at that date its design was practically what we now know as the Red Ensign or British Merchant Navy flag.

∘NO PROHIBITIONIST∘

Water was used to christen the U.S.S. Constitution on the first two attempts to launch her. She wouldn't budge. Then someone produced a bottle of Madeira, smashed it against her bow, and she slid gracefully into the stream.

• RECRUITING COSTS •

Up until as late as 1820, recruiting costs, including printing of posters, etc., were borne by the ship's commander.

• FIRST MARINES •

The first Marines were known as "The Duke of York and Albany's Maritime Regiment of Foot." In 1664 there were 1,200 of them.

• SALT JUNK •

Old Navy slang for corned beef. As junk was the product of old rope unrove to be used in caulking seams, one has only to imagine the stringy quality of the salt beef to discover there was more truth than poetry in the name.

• BY THE GREAT HORN SPOON •

Here is a sea-going oath that has been out of circulation so long that no man living has heard it used as such.

It is supposed to have reference to the Big Dipper, which some early mariners knew by the nickname of the "Great Horn Spoon." This name in turn referred to the days when feeding utensils were primitive and few. The wives and families of lowly people had small spoons carved from horn; the head of the house had a *large* horn spoon, which served the double purpose of helping the family from the common bowl, and by which he fed himself.

So lonely mariners far at sea, nostalgic for the comforts of their own hearths, remembered the simple table utensil and what it stood for, and called the starry constellation the "Great Horn Spoon."

· EIGHT BELLS ·

This measure of time originated in the days when a half-hour glass was used to tell off the four-hour watches. Each time the sand ran out, the ship's boy, whose job it was to reverse the glass, struck a bell to show he was attending to his business. Thus, eight times he turned the glass, and eight times struck the bell.

· DUTCH COURAGE ·

Slang-term for bolstered heroism. It was coined to describe a custom of the old Netherland's Navy, when it was common practise to serve schnapps or gin to the gun-crews of fighting ships prior to engaging in battle.

· IN MUFTI ·

Meaning dressed in "Civvies". The word is a corruption of the West Indies "MUF-TEE", which means civilian.

· BULLY-BOYS ·

Nickname for sailors in Colonial days. However, it did not refer to any bullying or bad-tempered characteristics, but to the "Bully-Beef", standard ship grub of the time.

• BLUE NOSE •

Remembering the winter climate of the regions adjacent to and including the island of Nova Scotia, the inference of the term "Blue Nose" as applied to Nova Scotians needs no further explanation.

• LARGEST SILVER CARGO •

Page Henry Morgan, Blackbeard, and Captain Kidd! What a haul for a pirate in search of treasure ships!

On one of her calls to the Port of Los Angeles the American President Liner SS *President Garfield* carried no less than ten million dollars in silver.

• UPSIDE DOWN HATCH COVERS •

Hatch covers left upside down on deck are harbingers of bad luck. Any good sailorman can tell you that. In the old days it was thought that such carelessness would give evil spirits a chance to sneak down and bewitch the cargo.

• STEAM SHIP •

It seems odd that the inventor of the first practical American steamboat should have been neither seaman nor engineer.

Robert Fulton was an artist of national reputation, and like many noted painters was also an inventor.

So engrossed did he become in his mechanical brain-child that he practically deserted his ART, and devoted his life to sea-going transportation.

• IN THE DOG~HOUSE •

If you think this is modern slang, please note
that it was coined in or around 1800. When
slaves were bringing big prices on the American
auction-blocks, it was the brutal custom to
pack them into every available niche aboard
the slave-ships. Even the officer's cabins were
filled with them, while the officers had to sleep
on the poop-deck in semi-cylindrical boxes, six
or seven feet long, and about thirty inches high.
These boxes were nicknamed "Dog-houses," and
because they were so horribly uncomfortable to
sleep in, the term "In the Dog-house" grew to
describe being in a tough spot.

• PAINT WASHERS •

Term of contempt, used by the hairy-chested sea-
dogs of wind-driven warships, to describe the young
"upstarts" who served aboard the new-fangled steam-
driven battle wagons.....and dared to call themselves
sailors.

FLYING~FISH SAILOR

Old Navy slang to differ-
entiate between a seaman
on duty in Asiatic waters,
and one in a Mediterranean
squadron. The latter was
known as a "Sou'Spainer."

• BLOODY •

A meaningless expletive
much in use among British
sailors. Originally a pious oath
"By Our Lady," rough seamens'
talk corrupted it to "By oor Leddy"
"B'oor Luddy," and finally,"Bloody."

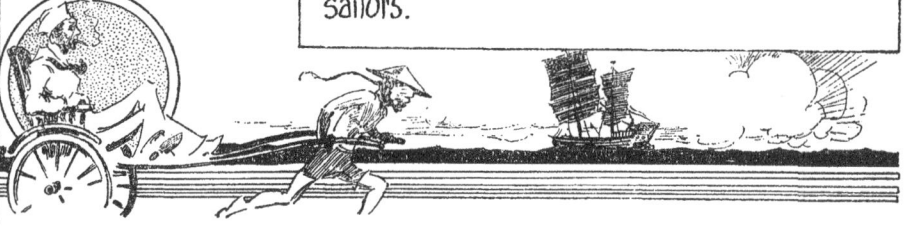

• CLIPPER •

The meaning of the term clipper bow is gleaned from the old English word "clip," to go at great speed.

• "CHINESE" GANGWAYS •

As the traditional direction for gangways to face is aft, any accommodation ladder which reverses the position and points forward is known as a Chinese gangway, meaning back to front.

• MARINE OFFICERS' SWORDS •

The design for the modern marine officer's sword is practically a duplicate of the weapon which was presented to William Eaton by the Bey of Tripoli, for his services in destroying the power of Barbary pirates.

• VANDALISM •

Give a dog a bad name and it sticks to him. The word vandalism (and all the useless and unnecessary destruction it implies) was coined for the acts of the semi-civilized pirates known as Vandals.

Yet history itself records how little the Vandals were given to wanton destruction.

Take the case of their sacking of Rome. True, they took all the gold out of the city after its capture, but they mistreated no one bodily; neither did they destroy civic buildings and homes.

In fact it would seem that many modern invasion armies with legal status might well have patterned their behavior after the methods of the Vandals, to the great good of everybody concerned.

WINDAS

· DEAD   MARINES ·

This slang-term for empty bottles was coined when, at a banquet aboard ship in his honor, the Duke of Clarence ordered the removal of the "empties," saying "Take away those dead marines." A major of marines asked haughtily "Why nickname empty bottles after *my* honorable corps?" "Because" replied the ready witted Duke, anxious to give no offence, "they are fine fellows who have nobly done their duty, and, if filled once more, would be willing to do so again."

· JOSS ·

Chinese name for a god. It is really a corruption of the Portugese "Dios" and was learned from Magellan's sailors when they visited the Orient.

· FIRST NAVY UNIFORMS ·

Until the year 1747 the dress of sailormen was very nondescript. At that time King George II ordered uniforms to be worn by all navy men, as a means of boosting their appearance and morale.

·ONCE *in a* BLUE MOON·

Believe it or not, this phrase is <u>not</u> just a whimsical saying used when expressing any unusual happening, but actually refers to a condition of the moon, which on certain rare occasions assumes a bluish color. It was last reported by the British Navy in 1883.

• SOME PAINT JOB •

The amount of paint consumed in dolling up the modern United States cruiser runs well into one hundred tons.

• FIRE CHIEFS •

This is a modern nickname aboard vessels in the Orient Trade for coolies who "ran amuck." They are known as "fire chiefs" because they invariably grab a fire-axe from the rack before sallying forth to wreak vengeance on the ship's company.

• RECORD "MOTHER-SHIP" •

This might well have been the nickname bestowed by contemporaries on the good ship *Independence* during her twenty-six years of faithful service on the North Atlantic run.

No fewer than fifteen hundred babies were born aboard the gallant packet, and it seems she brought each and every one of her precious charges safely to shore.

• CATHEADS •

The catheads were short heavy boomkins extending from the rail to port and starboard, near the bows.

Secured slightly aft and above the hawseholes, they were used for bringing anchors aweigh, and were found on most ships until the beginning of this century, when they became obsolete.

They got the name "catheads" because in very early times they were surmounted with carvings representing cat's heads, the cat in those days being a royal pet.

· BERSERK ·

This word, denoting ungovernable rage, was coined from a custom of the Vikings, who, to prove their courage in desperate fights, would tear off their shirts of mail, and fight half-naked. Hence the term "Ber-serk," or "Bare-sark," meaning literally "Bare of Shirt," as sark is the Norse word for shirt.

· DERRICKS ·

Ship's loading booms received their name from an enterprising hangman of Queen Elizabeth's reign. Mr. Derrick was a true artist, and invented a swinging beam for his gallows, with topping lift 'n' everything.

· PORTHOLES ·

Port holes were originally gunports. In early days no provision whatever was made for admitting air or light into the crew's quarters, which remained foul and gloomy until recent times.

· CAREEN ·

This method of cleaning or repairing ship's hulls, was named for the French word "carin", meaning "to turn over."

• A GREAT LIFE SAVER •

To Sir Arthur Roston, RNR, goes the distinction of being credited with having saved more lives at sea than any other modern mariner.

• THE "OVERLAND" COURSE •

Shades of the Pony Express and Buffalo Bill! Now comes official information, that by a series of connected canals and locks, it is possible for vessels of light draught to sail from Olean, New York, to Fort Benton, Nebraska.

• MONITOR STILL VISIBLE •

The old *Monitor*, which sank the Confederate *Merrimac* during the years that North and South were discussing the merits of the slave trade, can still be seen in the translucent halls of Davey Jones' locker. Wedged fast on the Diamond Shoals, where she sank in 1862, the battle-scarred old sea-dog clings steadfastly to her last station, as if defying Time and Fate to do their worst.

• A COIN FOR CHARON •

When burying a shipmate at sea, old-time sailormen would place a penny in the dead man's mouth before sewing him into his canvas shroud.

The coin was to pay for their deceased comrade's transportation across the river Styx, and would be collected by Old Man Charon for that purpose, when the corpse had been ferried across the water and delivered safely to the Better Land.

WINDAS

· BOOTLEGGER ·

This is a modern term with an old origin. Bootleggers got their name from smugglers in King George the Third's reign. The nickname derived from the smuggler's custom of hiding packages of valuables in their huge sea-boots when dodging His Majesty's coastguardsmen.

· HOG ~ YOKE ·

Old Navy slang for that indispensable instrument of navigation ....the sextant.

· SCOTTISH ENGINEERS ·

So strong is the tradition that Scotchmen are born marine engineers, that many folk maintain you can yell "Are you there, Mac?" into the engine-room of any ship afloat (regardless of nationality) and receive an affirmative answer.

· ONE CASUALTY ·

Amazing as it seems the American fleet in the Battle of Santiago lost but one sailor, while the Spanish Admiral Cerveras lost every vessel under his command and 500 men.

• CHARTS •

The first early English charts were known as sca-cards or sea-cards.

• FLEETS •

The word "fleet" is derived from the Spanish *flota*. From the same source is the word "flotilla."

• BRITISH ENSIGN •

The British ensign as we know it today received its nickname of UNION JACK from good QUEEN ANNE.

• HANDS OFF •

As punishment for pulling a knife during a fight with a shipmate, old Admiralty laws decreed that the offender should have his hand cut off.

• A "TOM SAWYER" TRAVERSE •

Mark Twain's remarkable book character had a habit of tinkering with unimportant little jobs, particularly when some task which he hated demanded his immediate attention.

With this thought in view it has become a shipboard colloquialism to designate as "making a Tom Sawyer traverse," the seaman who wastes time doing odds and ends, before applying himself to the job to which he has been assigned.

· YANKEE ·

The nickname of YANKEE was first applied to Americans by merchants of Holland. Because of the argumentative traits of certain American captains trading with the Netherlands, Dutchmen jeeringly called them "Yankers" (wranglers), and the name stuck...

· DAVITS ·

These devices for hoisting boats were named for their inventor, a Welshman named David, and given the Welsh pronunciation of that word viz. Davi<u>t</u>.

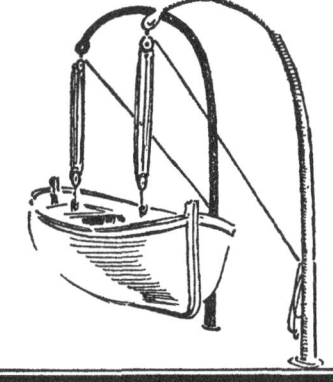

· LISTLESS ·

This good dictionary word, meaning dull or lifeless, was coined in the days of sail, when, under a good breeze, ships would list sharply to port or starboard. When there was no wind, they would ride without any list (listlessly) sluggishly, devoid of pep.

·LATITUDE and LONGITUDE·

Latitude and Longitude were first arrived at by astronomical observation, by Arabian ship-masters way back in the 15th. century.

• A LONG TIME AGO •

The first attempt to dig the Suez Canal was started by King Pharaoh Necho of Egypt in the year 610 B.C.

• CHITS •

The name "chit" for a note or voucher, was introduced into our language in the days of the East Indian Company.

A number of other Hindustani words were also added to our vocabulary through the medium of this old "Honorable John" line.

• GUN TURRETS •

Revolving turrets were invented twenty years before the Civil War, by T. R. Timby of New York when he was only nineteen years of age.

Used today on war tanks and aircraft as well as surface vessels, the turret has proved its worth in each succeeding war since its inception.

A LONG WAY FROM HER ELEMENT

Kaw Point Landing in Kansas is where the famous explorers Lewis and Clark once stopped, and where, later, flat-boats and river boats brought trade goods and pioneers to the Middle West.

Kaw Point came into prominence once again when shipyards in that section began turning out heavily armored deep-sea ships for foreign service.

If Hitler ever doubted America's industrial initiative, this should have given him food for thought, for Kaw Point is two thousand river miles from the nearest sea.

• SEA SHANTIES •

Sea chanteys (or shanties) received their name from an old custom of the negroes in the West Indies. When these men moved from one job to another, they would drag their shanties (huts) with them, what time one of their number would sit astride the hut and sing melodies to put a swing into the work of hauling. English sailors watching the manoevers forever afterwards associated the hut with the tunes, and gave the nickname of "sea-shanties" to all sea songs.

• LIBERTINE •

Today a word implying a person of easy morals, its origin meant nothing of loose living, but was the name for sailors on Roman ships of war.

•CHAIN CABLES•

"Old Ironsides" was one of the last American ships to use hemp cables for her anchors.   In 1812 chain cables were introduced, and quickly demonstrated their superiority over rope.

•KEEP A SHOT in the LOCKER•

Old navy slang which was the early equivalent of the modern "put a little bit away for a rainy day." It got its meaning from an admiralty order advising captains to "Keep always good reserve supplies in the shot-locker."

• DEEPEST WATER •

The deepest spot yet found in the ocean is near the Philippine Islands, where a sounding of thirty-five thousand four hundred feet has been reported.

• ENSIGNS AT HALF-MAST •

In the old NAVY, only officers of the rank of Captain or higher, rated the death-ceremony ritual of having the ensign flown at half-mast as a mark of respect at their passing.

• USS ALLIGATOR •

The Civil War submarine *Alligator* was built for the Union navy as a challenge to the Confederate ironclad *Merrimac*.

She was forty-six feet long and was propelled by sixteen oars, but lack of planning for an adequate supply of air for the crew while submerged made her an unsatisfactory experiment.

• HE NEVER WENT TO SEA •

A Confederate naval officer who caused the Union as much grief and trouble as any other individual during the Civil War was Asbury Harpending. He secured a commission in the Confederate navy, so that he would not be executed as a pirate if caught while attempting to seize a Pacific Mail steamer on the high seas, and deliver her cargo of gold to Jeff Davis.

Oddly enough, although Harpending plotted this and other sea exploits and was a duly accredited naval officer, he never went to sea in his life, other than as a passenger.

· SHANGHAIED ·

Just in case you didn't know it, the term "Shanghaied" originated in the Chinese port of Shanghai. Here, masters of American tea-clippers delayed for want of crews, would pay the Chinese owners of dives where drunken sailors were carousing, to slip drugs into the seamen's drinking glasses and hustle the unconscious sailors aboard the waiting ships.

· CLIPPER BOWS ·

The "Clipper" bow receives its name from the old English word "Clip" meaning "to be able to run at a fast pace."

· The LUCKY BAG ·

The Naval Academy's Year Book is named for certain lockers on old-time ships, wherein were placed all lost articles. Once a month it was a seaman's privilege to re-claim from said lockers such articles as he had lost during that period.

COPYRIGHT · C.W. WINDAS 1941 ·

· YACHTS ·

Now that so many yachts are being turned over to the Government by patriotic owners, it is worthy of note to discover that the first yachts were built by the Dutch. They were named for the Dutch word "Jagan" (to hunt) and were the equivalent of our modern coast-guard vessels.

• REVERSAL OF RANK •
AT FUNERALS

The reversal of the precedence of rank at naval and military funerals, whereby the seaman marches ahead of the commander, and the commander ahead of the admiral, is a practical demonstration of the teachings of humility as a cardinal virtue. It is based on a paragraph from the Bible which states . . . the first shall be last, and the last first.

• DRINK TOAST SEATED •

British naval officers have the privilege of remaining seated when drinking the toast "THE KING."

This is because when Charles the Second of England was rising to drink a toast while dining aboard a ship of the line, he bumped his head on one of the overhead beams.

The royal monarch took the accident in good part, laughing heartily with the rest of the guests; but he decreed that from that time on, officers drinking the royal toast could remain seated without incurring regal displeasure.

• WORLD'S GREATEST SEA-RACE •

The longest and greatest deepwater race known in history was between seven American and British tea-clippers towards the end of the last century. These handsome packets, set for the course which led them clear around the world, left Canton, China, within a few hours of each other.

Each taking her fair share of adverse weather and unusual hazards, they finally came booming on their way up the English Channel to the Port o' London; and the time difference between the first and last arrival was barely forty-eight hours.

∘ WHEN MY SHIP COMES IN ∘

This phrase, meaning, "If and when I make my fortune," was coined in the days when seafaring adventurers would send their fleets along the Mediterranean and African coasts in search of rich cargoes. First they would have to go to the money-lenders, in order to finance the venture. As it was impossible to set an exact date for the fleets return, they would sign documents promising to repay the loans "When my ships come in."

∘ SAILOR'S PANTS ∘ EXEMPT

You will be glad to know that a seaman's pants (and indeed his coat also) cannot be held for rent arrears. As late as 1938 a San Francisco landlord was haled into court and fined for breaking this law.

MEDIEVAL NAVAL ∘ ARTILLERY ∘

In strange contrast to the monster rifles in our modern turrets, is this medieval naval gun, which operated on the principle of a windmill. It was used to throw huge stones onto the enemy's decks.

RECEIVE PAY ∘ on CAPS ∘

A tradition of the British Navy is that enlisted men receive their pay on their cap tops. This was originally done so that all could see the amount paid, and any errors could be readily rectified.

• CARGO •

The name cargo is derived from the ancient Latin *carga*, meaning a load.

• THE SAILMAKER'S GUINEA •

Sailmakers in the British Navy still receive one guinea for every corpse they sew into a canvas shroud for burial at sea.

• DRUM MAJORS •

The smart gestures and prancing gait of the modern drum major (and his sister the majorette) are a relic of the days when seamen ashore after a naval victory were encouraged to march along shouting and waving their arms (and side-arms), leaping into the air to express exuberancy.

This was done to work up a little public enthusiasm for their government's war efforts.

• HMS FURIOUS •

The British aircraft carrier *Furious,* "grand old lady of the British Fleet," has hung up a record it would be hard to beat. Laid down in 1915 as a special cruiser, she played a gallant part in the first World War by destroying the Zeppelin Works at Tonden. Later converted into a carrier she took on the whole German airforce in the Norwegian campaign. After that she sailed one hundred and thirty thousand miles of blue water from the Equator to the regions of the North Pole; took part in over a dozen engagements from which she emerged victorious.

· BOTTLED UP ·

Today a term in naval par-
-lance, indicating that an enemy
fleet is so completely hemmed in
that it cannot manoeuver, the
phrase itself was coined to illus-
-trate the quaint custom of old
sailormen, who would carve little
models of ships on which they
had served, and put
them in bottles.

WINDAS

· CARRY ON ·

While the order "Carry
On" now means only
to proceed with any
duty, it was originally
a specific order not to
shorten sail, but to
carry on all canvas
the ship would stand
unless stress of bad
weather dictated
otherwise.

· BUCCANEER ·

Alas for the dreams of boyhood! The
original buccaneers were not romantic
pirates, but only a bunch of hard-working
swabs who made "Boucan" (smoked beef)
for a living. Hence Boucan or Buccan-eer.

· MIND YOUR P'S and Q'S ·

Nowadays a term meaning "Be on your best
behavior." In old days, sailors serving aboard
government ships could always get credit at the
waterfront taverns until pay-day. As they
would only pay for those drinks which were
marked up on the score-board, the tavern-keeper
had to be careful that no Pints or Quarts had been
omitted from the customer's list.

• AN "AMY" •

Amy is the nickname bestowed on aliens serving aboard British warships. It is derived from the French word *ami* meaning friend.

• TOOK THE WIND OUT • OF HIS SAILS

Here's an expression by which we describe besting an opponent in some argument.

Originally it was a maneuver by which one vessel would pass close to windward of another, thereby blanketing the breeze from the other's canvas and making him lose way.

The term itself is picturesque and its meaning essentially practical.

• AMAZON RIVER •

The longest river in the Americas was named by Orellana, the famous navigator.

According to his own statements, he and his crew were exploring the strange new land they had come upon, when suddenly they were attacked from the banks of the river by huge and warlike women.

They were expert in the use of bow and arrow; handled their spears like veteran pikemen. They were led by one whose prowess was only exceeded by her great beauty. From early morn they fought, harrying Orellana's band with desperate attack after attack.

Then as the sun sank low in the west like a blood red ball of fire, these fierce and fantastic females disappeared into the jungle and troubled Orellana no more.

A very interesting story: there is only one slight discrepancy in this historic record. Nobody had ever seen or heard of the women before . . . nobody has ever seen or heard of them since.

But it's a wonderful name for a wonderful river anyway.

WINDAS

· KNOTS ·

To ascertain the speed of his vessel, a British commander had knots tied at regular intervals in a coil of rope. The rope was then bent onto a log, and the log hove overboard. With an hour-glass, he timed each knot as it disappeared over the taffrail ___ thus originating the custom of telling off a ship's speed by knots instead of miles.

· HALLIARDS ·

Originally an order "Haul Yards", these two words were corrupted into one which now designates any lines used for hoisting sails, flags etc.

· JACK TAR ·

International nickname for government sailors, because of the custom among old navy men of giving their work-clothes a light coating of tar to waterproof them.

· THE CUT of HIS JIB ·

An everyday phrase referring to a man's general appearance. The term derives from the days when it was possible to distinguish French vessels at a distance because of their two small jibs, and a British ship by her huge single jib.

• THE O.A.O. •

Any midshipman from Annapolis can tell you that this is the proper and correct term to use when speaking of the "best girl."

It means she is the ONE AND ONLY.

• THE SWORD-HILT CROSS •

In medieval days every Christian knight had a cross embossed on the hilt of his sword as a solemn token that he would keep the faith.

To this day, the dirk carried by British midshipmen carries on its hilt the emblem of religious fealty.

• SLOP CHEST •

The slop chest was a locker carried on deep-sea ships from which the master of the vessel would disburse clothes to needy seamen during the long voyage . . . at a very excellent profit. The word slop **is a** corruption of the old English *Sloppes*, meaning breeches or trousers.

• FINDING AN "ANGEL" •

Here is a theatrical term with a nautical origin. It means the gentle art of finding a sucker with more sentiment than business sense, who will provide the funds necessary to produce a "sure-hit" show.

It had its origin in the fact that one Luis de SANT-ANGEL financed the westward voyage of Columbus which resulted in the discovery of America.

From the latter half of the former gentleman's name the term was coined.

· SIRENS ·

Ships' foghorns got their name from the women of Greek mythology known as SIRENS. Beautiful but cruel, they used their marvellous voices to lure sailors to destruction. Ulysses outwitted them by having himself bound to the mast, while his seamen, their ears plugged so that they could not hear the enticing voices of the women, rowed him safely past the danger zone.

WINDAS

FOUNDER of NAVAL POWER ·

The first known to recognise the importance of maritime power was King Minos of Crete. He established the first navy in 3,000 B.C.

· CUTTING A DIDO ·

This term to describe some smart-aleck cutting monkey-shines, was coined because the commander of H.M.S. DIDO used to delight showing off the superior sailing qualities of his brig by sailing in circles around the slow moving old ships-of-the-Line.

·BOOBY·

Here is a perfectly legitimate dictionary word used to describe a foolish person. It gets its meaning from the crazy antics of the booby bird.

• THE OLD MAN •

Captain, as applied to the Merchant Service, is a courtesy title only. His official rank is that of Master Mariner . . . and he's generally called the "Old Man."

• RAILROAD PANTS •

Nickname for the rows of braid down the outside seams of a flag officer's dress uniform trousers. The English equivalent is "Lightning Conductor" pants.

• SKYLARKING •

Here is another good dictionary word which grew from a nautical slang term.

Skylarking was first coined to express the fun enjoyed by robust young seamen who would scramble to the fighting-tops of battleships, and descend to the decks by sliding down the backstays.

It was a sort of "follow-my-leader" game, and called for lots of nerve and stamina.

• RUNNING HER EASTING DOWN •

This old term refers to a particular area of the globe between South Africa and Australia. Vessels bound south to Melbourne or Sydney would round the Cape of Good Hope, and bear eastward on the long haul to Leeuwin, the southwestern corner of the land "down under." They sailed before the ever-prevailing trade winds in that quarter, and spoke of this as "running her easting down."

· RAISING THE WIND ·

This popular slang term for raising funds for some specific purpose dates back to early days, when a shipmaster would go to a witch or fortune-teller and pay big money for the old harridan's assurance that good winds would surely drive his vessel for the entire voyage and bring her safely back home.

· FIRST U.S. NAVAL COMMISSION

The first U.S. Naval commission was that given to Captain Samuel Nicols of the U.S. Marines by the Continental Congress Nov. 28th. 1775 .

· LIFE ~ BUOY ·

The life-buoy got its name by the simple expedient of dropping the last three letters from the word "buoyant."

· SHROUDS ·

The side and backstays of a mast were called "shrouds" because in early times the quality of these standing ropes was so poor, that the vast number which had to be used for strength shrouded (hid) the mast from view .

• ADMIRALTY "HAM" •

British navy nickname for canned fish.

• BAMBOOZLE •

Here's a perfectly good dictionary word which had its origin in early day colloquialism. It meant the act of deceiving passing vessels as to your nationality, by flying some ensign other than your own; a common practice of pirates.

• ADMIRAL'S "EIGHTH" •

This term refers to the prize monies paid to Admirals for all ships captured by the fleet under their command, whether the flag officer himself was present or not at the time of their taking.

One eighth of the value of the ship allowed by the prize courts was his . . . and a very tidy percentage, we'd say.

• FIRST ATLANTIC CROSSING • BY STEAM

The first credited North Atlantic passage by a steamship was made by the *Royal William* of Canada.

It raised great joy among sailing ship "die-hards," because the vessel took just as long on the run as many of the fast windjammers of the day.

But men of vision hailed the voyage with acclaim, for it proved the practicability of steam for deepwater work.

The *Royal William* rang the death-knell for sail and marked the first great step forward in ocean travel.

· ADMIRAL ·

It seems a far cry from a Moorish chief to a senior ranking officer in the Navy. Yet a Moorish chief is an Emir, and chief of all the chiefs is the EMIR-AL, from which we get our English word "Admiral."

WINDAS

·FLEUR de LIS·

The fleur de lis, which points the North on all compass-cards, was put there originally by the 14th century Neapolitan pilot who designed the card; it was a tribute to his king, who was of the house of Bourbon.

N

· SON of a GUN ·

This term dates back to when men of certain ratings, including gunners and gunners' mates, were allowed to take their wives along to sea with them. If a boy was born on the voyage, he was half-humorously, half-contemptuously referred to as "a son of a gun."

· IN THE BRIG ·

Because Admiral Nelson once assigned a small brig to carry captives taken in one of his naval engagements, and because his seamen ever afterwards associated that vessel with prisoners, the name 'brig' became sailor's universal slang for JAIL.

• SCUTTLE •

This is practically the original form of the old Anglo-Saxon word which means "Hole."

• THE "HUNGER" LINES •

This term had no connection with charity bread lines.

It referred to shipping companies whose vessels were notorious for poor food, hard work, and little pay.

• THE PHAROS LIGHT •

The great Pharos lighthouse at Alexandria was three hundred and seventy feet high and was one of the wonders of the world.

It was built, so say the early chroniclers . . . "so that all nyght, mariners dryven before ye raging tempest could make greate comforte from its guyding rays."

Which they did, from 274 B.C. until the year of Our Lord 1326. And that's a good record for stability in any man's language.

• HOUSE-FLAG'S ORIGIN •

The flags of the various shipping companies are known as "house-flags" because the device borne on such flags is the insignia of the company or house which operates the ships.

Their origin dates back to medieval days, when Crusaders, off to the Holy Wars, each carried on his ship a banner showing the crest or coat-of-arms of the house or family to which he belonged.

The little enamelled replica of the companies' flag which modern merchant marine officers wear in the oak-leaf-framed badges on their caps, make a bright touch of color as if reminding us of their brave origin.

WINDAS

• BRINGING THE WAR CLOSE HOME •

To the JACOB JONES belongs the distinction of being the first U.S. destroyer ever to be sunk by enemy action in <u>home</u> <u>waters</u>.

• FIRST U.S. OCEAN MAIL •

In 1847 Congress ordered built a fleet of steamers under the great Commodore Perry to carry the first U.S. Government Ocean Mail Service

BLUFF OLD SEA-DOG DIES

In view of the battle now raging in the Crimea, it is of more than passing interest to note that there died recently in Brisbane Australia, an ancient mariner named Charles Longden who served in the Crimea War of 1854. He was 105 years of age.

• BLUFF •

And speaking of the word "bluff" (meaning stout or hearty) the word was coined from the general appearance of stoutness or strength of old ships' bows.

• AYE, AYE •

This affirmative expression is generally supposed to be a corruption of the words Yea, yea.

The claim is advanced that Cockney accents changed the Yea to Yi, and from there it was a simple transition to Aye.

• TAFFRAIL •

Some maintain that the word taffrail is a combination of three words, namely:—the after rail. Others claim it is a misspelling of the Dutch word *Tafareel* which meant painting on the stern. Take your choice; they are both practical.

• RED BULWARKS •

As late as John Paul Jones' day, the decks and also the inboard side of the battleships' bulwarks were painted red.

This was done so that young seamen, going into action for the first time, would not be dismayed by the sight of blood splashed around in the height of the battle.

• BODY AND SOUL LASHING •

When the weather was particularly dirty and the decks awash with heavy seas, sailormen would secure lanyards or small stuff around their waists, the bottoms of their trousers, and the cuffs of their jackets, to keep the water from drenching their underwear and making them feel more miserable (if that was possible) than ever. This was called "bending on the body and soul lashings."

WINDAS

· FORECASTLE ·

This name (fo'c'stle to you) for the crew's quarters, is a relic of the days when huge wooden castles actually were built on the fore and after ends of ships, from which fightingmen could throw spears, arrows, stones etc., onto the decks of an enemy.

·EXECUTION DOCK·

A familiar phrase today for any place of legal lethal dispatch. However, there was actually a place by that name, situated near Blackwall (England) next to the East India Docks on the Thames.

· QUARANTINE ·

This term for medical detention derives its name from the French "Quarant," meaning forty. The first known case of isolating a ship for reasons of plague, was at Marseilles, and the vessel was held for FORTY days... hence the name..

· LIMEY ·

Because it was practically impossible to carry fresh fruits and vegetables on protrated voyages years ago, British Parliament decreed that each sailor must drink a pint of limejuice daily as a preventative against scurvy. Thus came the nickname for British ships, and Britishers in general.

• PAINTERS •

The light line secured to a small boat's stempost receives its name from the French word "Peyntours," meaning a noose or bight.

• PADDY'S MILESTONE •

Ailsa Craig, the island near the entrance to Glasgow harbor, became known as Paddy's Milestone because of the large number of Irishmen who crossed to Scotland.

These sons of Erin were cheered by the sight of the nearness of their goal when they sighted the island, and gave it the affectionate nickname above noted.

• THE LOCKED SEA-CHEST •

Old sailormen were very jealous of their simple honesty. If a greenhorn on his first voyage should carefully lock his sea-chest for safe-keeping, some older member of the crew would take him aside and gently but firmly point out to him the error of his ways; for a locked sea-chest was a personal insult to every seaman in the fo'c'stle.

• THE "BLOODY" FORTIES •

The "Bloody Forties" was an association of the rip-roarin' sailormen who manned the clipper ships running between New York and and Liverpool.

They called themselves by this name because the clippers used the sea-lanes in the circle of North Latitude 40 degrees, when driving on their east to west passage.

WINDAE

The FIRST SUBMARINE

In the present war in the East, Dutchmen are really living up to the best traditions of their race. Particularly is this true of their submarine successes. And rightly so, for it was Cornelius Van Drebel, a famous Dutch scientist, who invented the submersible in 1622.

The worthy Hollander interested King James of England in his ingenious product, and actually sailed it both on the surface and underneath the Thames River.

The craft was built of wood and was propelled by six oars.

The 'HUNGRY HUNDRED'

In the gay '90's, this was the nickname bestowed on the first group of Royal Navy Reserve officers to be assigned to duty with the fleet; the significance of the term referred to the insignificance of their pay.

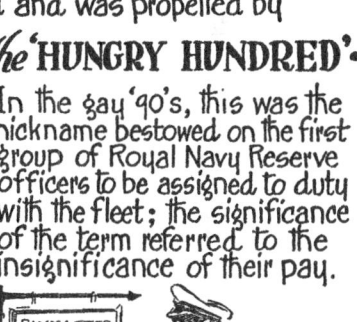

DAGGER~RAMMER~REES

This is the awesome pronunciation given the name of Diego Ramirez Island by unscholarly seamen in the early days; and "Dagger-Rammer-Rees" it remains to this day in sea-going circles.

The SAGRES

The beautiful Portugese naval training ship which recently visited the U.S. was named for the first Naval Academy ever to be established.
This institution was founded at Sagres Portugal, by Henry the Navigator in 1415.

• WATER BEWITCHED •

Old time shipboard slang for weak tea.

• CROSSING THE "T" •

Crossing the "T" is the dream of every fighting naval officer's heart.
To catch the enemy's ships in line of column, so that our broadsides can concentrate on his leading ships, which, head on to us, are deprived of the full use of their main batteries . . . that is a maneuver which brings joy to our hearts and grief to the enemy.

• JERSEY •

That woolen sweater you wear is supposed to have been named for one of the Channel Islands, the island of Jersey.

To say that the people of this island were actually the first to make such a garment would be overstating a fact, but they get the credit for it anyway, and with the credit goes the name.

• KEEP YOUR WEATHER-EYE • LIFTING

With the passing of sail, the admonition to "keep your weather-eye lifting" has evolved into a bit of cautionary advice to watch your step, or to take good care of yourself.

In the old days, however, it was a very pertinent order. It meant that the helmsman must keep his eye on the weather-leaches, and so handle his steering that no wind was spilled from the sail. This called for constant vigilance as a quivering leach was a prelude to loss of speed.

WINDAS

ARCHIMEDES' WAR MACHINE·

Strangely enough, though he was neither sailor nor soldier, Archimedes invented the earliest machines of war which proved so devastating to the Roman fleet at Syracuse. They were huge grapnels which lifted a vessel's bows from the water, smashing and then dropping her under stern first.

·DISCHARGING RIFLES *at* SUNSET·

The custom of Royal Marine sentries in discharging their rifles at sunset is a relic of ancient times, when small arms were fired to prevent possible misfiring later on, owing to dampened priming. Each new sentry, of course, used fresh charges and priming.

·BOX *the* COMPASS·

This term for reciting the points of the compass is claimed by some to have originated from the Spanish word "Boxar", meaning "to sail around."

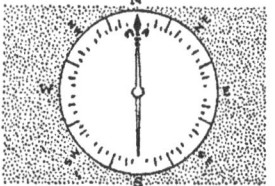

·WEST COAST MOUNTAINS·

Three famous Pacific coast mountains were named for British Admirals; Hood, Ranier and Helena.

• TRIM THE DISH •

Just an order in a small boat for the occupants to so dispose themselves that the little vessel will sail on a fairly even keel.

• GALLEY •

The most logical explanation for ship's kitchens being called galleys, is the one which maintains the word is a corruption of "gallery."

Ancient mariners cooked their simple meals on a brick or stone gallery laid amidships.

• DONKEY'S BREAKFAST •

To the old-time sailorman, inured to sleeping in hammocks or on bare bunk boards, a mattress was the all time high in luxury.

Even though it was stuffed with straw, it marked a big step forward in the march of progress, and because the first sea-going mattresses were invariably filled with such material, the name "donkey's breakfast" became a synonym for seamen's beds.

• S.O.S. •

No, Sally. S.O.S., the universal wireless signal for ships in distress, does not stand for "Save our Souls."

That idea may have been the dream-child of some romantic publicity man, but wireless operators promptly jeered the idea into oblivion. They explained that the letters S.O.S. were just a quick and compelling combination to command instant attention.

The letters themselves have no hidden meaning.

WINDAS

· NO QUARTER ·

This term, indicative of a fight to the death, gathers its meaning from the reverse of "Giving Quarter," an old custom by which officers, upon surrender, could save their lives by paying a ransom of "One Quarter of their year's pay."

· LEATHER~NECKS ·

U.S. Marines earned this name in 1812, when stiff leather bands were sewn into their collars to ward off sword strokes.

SALUTING *the* QUARTER DECK ·

A relic of Roman days, when images of the gods were housed in that section of the ship, and were paid homage by everyone as they came aboard.

· KEEP YOUR SHIRT ON ·

Slang for "Don't get fighting mad," it refers to the act of the ancient Viking berserkers, who in fury would tear off their shirts of mail, and fight half naked.

• PURSER •

The ship's purser was originally known as the "Burser," the man who had his fingers on the vessel's money-bags and paid bills, wages, etc. From this same derivation we get also the word "disbursement."

• SHOVE OFF •

Remembering the action contingent with a small boat leaving a ship's side, it is only natural that Jack should refer to any and all leave-takings as "shoving off."

• HOOKER •

While this slang term is used more or less scathingly by modern seamen when referring to some old ship of their acquaintance, it originally described a certain line of small vessels trading between British ports and the Hook of Holland.

There did not seem to be any intent of ridicule when the ships first received the name of "Hookers," but this note of disparagement has crept into the modern use of the word.

• BEACHCOMBER •

This term of opprobrium for a ne'er-do-well or loafer along the waterfront originated in the islands of the South Pacific. It was terse and true, for it described the outcasts of the island ports, who, lost to all ambition, actually preferred to comb the beaches for a scant subsistence rather than earn a decent living by good honest work.

WINDAS

· STARBOARD SIDE ·

Because the Vikings shipped their star (steering) oar on the right hand side of their vessels, and called the side of a ship its "board", the right hand side of vessels has ever since been designated as the "star-board" side.

· DEAD RECKONING ·

This navigation term was originally spelled "ded" (the abbreviation for deducted) reckoning. An unscholarly British shipmaster thought the "a" had been ommitted, so inserted it. Ever since then, even the officially printed forms spell it "dead" reckoning.

· PILOT ·

From two Dutch words "Peil" (to mark with pegs) and "Loth" (lead). Strangely enough the name was conferred on persons who could navigate a vessel into port without the use of the "Peil-loth" (lead line).

· LOG BOOK ·

As early ship's records were inscribed on shingles (cut from logs) and hinged so that they opened like a book, the name "Log-book" was logical and lasted to this day.

• BUOY •

The creation of this word was simple; knock off the last three letters of the word "buoyant" and there you have it and what it means.

• TYPHOON •

To the Chinese we are indebted for this addition to our nautical vocabulary. The very word breathes all the sinister implications associated with such a devastating disturbance. Typhoon means "the Mother of Winds."

• GINGERBREAD •

Decorations on a ship's stem or stern are always referred to as gingerbread. This is because for countless generations boyhood memories carry back to the delectable cookies made of gingerbread and covered with vari-colored scrolls of sugar. Which only goes to show that sailormen never grow up.

• TOPS'LS UNDER •

Here is a colorful term much in vogue among Old Navy men. It paints the atmosphere of the day in broad sure strokes. One visions the rolling sea with scudding clouds above. Far astern is a ship we have been racing. Gradually she disappears from view. The horizon engulfs her hull, her main courses. Even as we look, her topsails dip below the line, leaving only the white flash of her t'gallans and royals to mark her passing. We have left her "tops'ls under," meaning in the picturesque vernacular of the day that we have won the race.

· SWIFT JUSTICE ·

By old British Naval Law, any seaman who was found guilty of having murdered a messmate aboard ship was forthwith tied to the corpse and flung overboard.

· FIRST NAVAL COMMISSION ·

The first commission granted to a U.S. Naval officer was to Captain Hopley Yeaton, March 21st 1791. He was master of a revenue cutter, then the sole maritime defense of our young republic.

· IT'S AN ILL WIND.... ·

The old saying "It's an ill wind that blows nobody any good," was originally exclusively nautical, and meant that no matter which way the wind blew, some ship must surely profit from its direction.

· CRUISERS ·

The word cruiser is derived from the Mediterranean word CRUSAL· meaning "fast."

• TOPS'L BUSTER •

Old Navy slang for a howling gale. Its inference is too obvious to call for any explanation.

• ON THE BEACH •

A simple but wonderfully descriptive way in which a sailorman voices the fact that he is not only out of his element, but worse still, out of a job.

• LYING ON OARS •

This salute to a passing superior officer's launch or boat is paralleled by other forms of salutation, such as stopping a launch's engine, or letting fly the sheets of a sail.

They are all official methods of acknowledging the presence of a ranking officer.

• TUMBLE-HOME •

Tumble-home in a ship's construction is often thought to be so designed as an extra measure of strength or sea-worthiness.

As a matter of fact, tumble-home was first invented to beat Suez Canal toll charges, which were based on a formula of Length by Depth by half the Deck-Beam.

As the deck-beam was reduced at least twenty-five per cent in tumble-home construction, the shipowner saved many a pretty penny until the Suez formula was revised.

It may have been mere coincidence, but oddly enough, tumble-home construction lost its popularity almost simultaneously with the revision.

WINDAE

• DIPPING THE FLAG •

Dipping the flag is a survival of a very old custom when merchant ships were required to clew up all their canvas and wait until the adjacent man-o'-war either sent a boat off to inspect their papers or signalled them to proceed. The flag salute was later adopted as a time saver.

• FOR BLASPHEMY •

Despite Hollywood and popular opinion, cuss words have always been frowned upon aboard ship. In Queen Elizabeth's reign, blasphemy was punished by burning the offender's tongue with a hot iron.

• BO'SUN'S PIPE •

The present form of the bo'sun's pipe is actually a facsimile of that taken from the body of the infamous pirate Andrew Barton by Admiral Lord Howard.

• PALE ALE •

Old Navy slang for a drink from the scuttle butt.

• MOORINGS •

Here is a word supplied by the Netherlands. Moorings is from the Dutch word "Marren," meaning "to tie up."

• HE KNOWS HIS 5 L'S •

A tribute from one seaman to another shipmate's knowledge; meaning, "He knows all about latitude, longitude, lights, log, and lead."

• CRETE •

King Minos of Crete was the first to establish a regularly organized fleet, which he did in 1460 B.C. Thus to him goes the credit of being the Father of Naval Power.

• POSTED AT LLOYDS •

From Lloyds little coffee shop in earlier London, grew the world's greatest marine insurance company. From such humble quarters where masters and merchants congregated to discuss freights, rose the massive establishment now housing the documents relative to the shipping of the world.

Outside its doors is a bulletin board, a grim advertisement of ships that are overdue and must (after a certain lapse of time) be considered lost.

A million tired eyes have read the names of the gallant vessels which have never come home; a million aching hearts have turned away in sorrow for a ship that was "Posted at Lloyds!"

• SAILORETTE •

Just in case you should think it a modern innovation to grant commissions and ratings in the navy to the weaker (?) sex, remember that Mary Ann Talbot received a pension of Twenty pounds per annum for wounds received in action many years before Nelson's time. Also note that Ann Johnson died at the Battle of Copenhagen while serving as a member of a British gun crew.

• SHACKLE •

This word is derived from the old Anglo-Saxon "sceacul," meaning a link of chain.

• MEGAPHONE •

..... and if you think this is a modern instrument, be advised that Alexander the Great used one 335 years B.C.

•OAK LEAVES•

Oak leaves are used in insignia as a tribute to the memory of the staunch ships of oak in the good old days of sail.

• REAR ADMIRAL •

The title of Rear Admiral was first given to divisional commanders of reserve fleets. Hence the inference of being in Reserve, or "IN the Rear."

• COIN UNDER MAST-STEP •

From time immemorial, a small coin has been placed in the mast-step of a newly launched vessel, to propitiate the gods and bring good luck.

• SCREWS VERSUS PADDLES •

It is about one hundred years ago since screws were adopted as the superior driving force for battleships. In 1839 the British Admiralty ordered a tug-of-war between HMS *Alert* (a screw steamer) and HMS *Rattler* (a paddle wheeler). Made fast to each other stern to stern the ships steamed in opposite directions, and the *Alert* proved victorious.

• THE LUCKY BAG •

The so-called lucky bag was really a huge locker in which articles lost aboard ship were deposited. Once a month these articles were produced and handed back to their respective owners.

But there was a catch to it.

Each lucky recipient of a lost article was then given three strokes from the cat-o'-nine tails to teach him not to lose anything again.

WINDAS

· THE FIRST CHRONOMETER ·

The first chronometer was made and demonstrated aboard H.M.S. CENTURION, in 1736. Sir Isaac Newton declared the making of an instrument of such precision to be impossible. But the CENTURION'S carpenter built one of wood..... and it actually worked successfully. From that time on, longitude could be reckoned accurately.

· WHITE RATS ·

Slang term in the old Navy for men who would endeavor to curry favor with the officers by carrying tales aft.

· VIKING ·

The correct pronunciation of this word is "Veek-ing," and refers to those wild sea robbers who laid in wait for their victims in the viks (veeks) or bays of Norway.

· BOWSPRIT ·

While the jib-booms of yachts and other small craft are often called bowsprits, the term is lubberly, as the bowsprit is an integral part of a vessel's hull, to which the figurehead is secured.

• A "SPOTTED COW" •

This was a nickname for German ships registered at Hamburg. The term was coined for Simon of Utrecht, Lord of Hamburg, whose banner bore the device of a dappled bull.

• SEA-LAWYER •

Not a member of the legal fraternity by any means, but a surly fellow who is forever arguing about anything and everything aboard ship, with a view to getting out of scrapes (and more particularly) out of work.

• CHAIN STORE ORIGIN •

Clipper ships were responsible for the beginning of chain stores. These fast vessels brought home their cargoes of tea in such quantities, that brokers cut tea prices to almost nothing. In 1859, therefore, the owner of one big tea cargo decided to sell his merchandise through a number of stores he opened for that purpose.

• THE SUPERSTITION OF FRIDAY •

The reluctance of seamen to sail on a Friday reached such proportions, that many years ago the British government decided to take strong measures to prove the fallacy of the superstition.

They laid the keel of a new vessel on Friday, launched her on a Friday, named her HMS *Friday*. Then they placed her in command of one Captain Friday, and sent her to sea on Friday. The scheme worked well, and had only one drawback . . . neither ship nor crew were ever heard of again.

WINDA3

FIRST 'SUB' VICTIM

The first recorded victim of a submarine was the U.S.S. HOUSATONIA, attacked by the Confederates while off Charleston, S.C., on February 17th. 1864.

WE DEMANDS OUR SALT 'ORSE!

• BANIAN DAYS •

Years ago, when as a measure of economy, two meatless days per week for seamen were intro- duced, sailors sneered at them as "Banian Days", naming them for the Hindu Banians, a tribe whose religion prohibits any diet other than vegetarian.

• A 24~CENT 'PEEVE' •

When King Manuel of Portugal refused to raise the famous Magellan's army pay from $2.^{25} to $2.^{49} per month, that irate officer forsook his military career and turned navigator.

LATEEN SAIL

This name for a triangular mains'l was originally LATIN sail, so called to designate the rig of Mediterraean type vessels.

• DONKEY-ENGINE •

The donkey-engine was so named because the engine replaced the animal as a motive power on the whip-like contrivances used for hoisting heavy loads.

• A "LONG" SHOT •

Here's a modern gambling term with an old nautical origin. Because ships' guns in early days were very inaccurate except at close quarters, it was only an extremely lucky shot that would hit the mark at any great distance, hence the inference of "luck" in the gambling term.

• A MODEL THAT FIRED • A KING'S SALUTE

The U. S. Naval Academy owns a model of the French three-decker *Ville de Paris* with a very historic and unique background. At Dresden in 1814, Alexander the First of Russia was given a 120-gun salute from the little brass cannon peeping from the model's gunports.

• CHAPLAIN •

The "reverends" of the fleet are called chaplains because Saint Martin of blessed memory divided his coat with a poor beggar one bleak and bitter day near Amiens. It is recorded that the coat remained miraculously preserved and finally became a holy banner for the King of France.

It was hung in an oratory that was named "the chapelle," and the custodian charged with its safekeeping was called the chaplain.

• OLD "STICK~IN~THE~MUD" •

This term, describing a person who is non-progressive or of no account, had a gruesome origin. It dates back to the days when pirates were hanged at the edge of the Thames tide-waters. When dead the felon was buried in the mud..... "so that forever none might find his foul body, nor account for his soul at the Ressurection."

• SANTIAGO •

This well known South American city was named for the Portugese patron Saint Jago, but with the passing of time the name has been corrupted by illiterate seafarers, both in the spelling and pronunciation.

•The BRASS MONKEY•

International nickname for the very dignified golden lion on the crimson field of the Cunard Steamship Cº's handsome house-flag.

• ROVER •

While this name is now generally applied to buccaneers, it was originally the trade name for the hardworking brotherhood of riggers. These skilled men became known as ROVERS because they traveled from ship-yard to shipyard as their jobs demanded, much as our modern fruit pickers seek work.

• CREW •

We have the word "crew" from the old Norse *acrue*, meaning to gather; and from the same source also the word "recruit."

• WART •

This is the scathing nickname applied to British naval cadets. It is used more in sorrow than anger, on the international theory prevailing in naval circles, that youngsters aboard ship must be constantly reminded how utterly insignificant and useless they are to their superiors.

• GREASY LUCK •

Wishing you "greasy luck" was the old-time whaler's equivalent for bon voyage and good fortune.

It had reference, of course, to the number of barrels of whale oil he hoped you would secure, for your efforts on the whaling grounds.

• THE PRECISION OF • CHRONOMETERS

Ships' timepieces are really marvels of precision. The chronometer is warranted to run consistently; that is, if it loses or gains a few seconds each day this is unimportant, provided it does the same thing every day.

But a clock which lost or gained irregularly would be a source of great grief to a navigator.

However, to such a high peak of perfection has the modern chronometer been brought, that many have been known to lose or gain only a few seconds during an entire voyage to Australia and back.

The SHIP that CHANGED a RIVER'S NAME

The Oregon River's name was changed to that of Columbia River in honor of the U.S.S. Columbia, when that vessel called there in 1788 while on her famous voyage around the world.

· THE JINX CAT ·

Page Hollywood, here's a new "Oscar." Oscar is the black cat which was rescued by British sailors when they sank the battleship "Bismarck." Later, Oscar was saved from the British destroyer "Cossack" when that vessel was torpedoed. Again Oscar was jerked from a watery grave when the plane-carrier "Ark Royal" was lost.
Oscar now lives at Gibraltar..... so keep your fingers crossed.

·KNOW YOUR SPUDS·

Contrary to popular belief, potatoes were not first introduced into Europe by Sir Walter Raleigh, but by that doughty seaman the great Admiral Drake.

TOUGH LUCK for "HARPIES"

There is now a bank in Glasgow, Scotland, which opens special accounts for seamen under a plan by which they may draw only enough each day for living expenses. This is to discourage waterfront women from reaping a harvest on sailor's pay days.

• THE ROUND-BOTTOMED CHEST •

Just old Navy slang for a sailor's carry-all . . . his dunnage-bag.

• FIRST SEA STORY •

The record of the first tale of the sea is dated 2500 B.C. It is written on papyrus, and is housed in the British Museum. It recounts the terrific struggle between a doughty seaman of the day and a sea-serpent.

• SLIPPED HIS CABLE •

In the jargon of deep-water men, a messmate who has died has "slipped his cable."

The meaning of the phraseology lies in the fact that only in the direst emergency, such as escaping an enemy in a hurry, does a vessel willingly leave cable and anchor on the sea-floor when she departs her anchorage. Even more to the point, she never expects to return to that port.

• KEEL-HAUL •

Keel-hauling was a brutal punishment inflicted on seamen guilty of mutiny or some other high crime, in the "good old days" of sail. It practically amounted to a death sentence, for the chances of recovery after the ordeal were slight.

The culprit was fastened to a line which had been passed beneath the vessel's keel. He was then dragged under the water on the starboard side of the ship, hauled along the barnacle-encrusted bottom and hoisted up and onto the deck on the port side.

If the barnacles didn't cut him to pieces, and if he hadn't been drowned in the process of operation, he was considered to have paid for his crime and was free.

But as we said before, his chances of recovery were mighty slim.

WINDAS

· BLOCKADE~RUNNER "EDINA" ·

Southerners please note! The old s.s. Edina, famous blockade-runner which dodged so many Union warships during the Civil War, is still afloat and in service. As a boy, the illustrator of this page sailed in her on many a cruise between Port Melbourne and Geelong, Australia. Like Johnny Walker she is "still going strong!"

What! NO KISSES ?

Because they might possibly be used as code words, the traditional little crosses which every sailorman has used in his love letters, must now be eliminated by British seamen, according to Admiralty orders.

· COPYRIGHT · C·W·WINDAS · 1941 ·

· ENSIGN ·

This title dates back to when privileged squires carried the banners of their lords and masters into battle. Later, these squires became known by the name of the banner (the ensign) itself.

·SPIRITUAL *and* PHYSICAL LIGHT·

The Cordovan Lighthouse near Bordeaux, France, was once a church. It is reputedly the oldest maritime light in existence.

• DIFFERENT SHIPS... •
DIFFERENT LONG SPLICES

An old Navy colloquialism meaning that there is more than one side to an argument, and more than one way of doing a shipshape job. It is a philosophy of tolerance everyone would do well to practice.

• ANCHOR WATCH •

Anchor watch was originally stood *only* when the ship was tied up in dock and her anchors stowed on deck. Then a watch was posted, "Lest," says the serious chronicler of early days, "some miscreants from ye other ships about, steal ye anchors while theye (the crew) sleepe."

• KIDS •

The shallow wooden (or metal) vessels used in carrying food from the galley to the mess table received their name from the boys whose job it was to help the cook aboard old time ships by waiting on the seamen at mealtime. The boys were "kids" in the vernacular of the day, and the wooden trays were named after them.

• CAME IN THROUGH THE •
HAWSE-HOLE

When a sailor has risen from the rank of ordinary seaman and attained the rank of master or captain, he is described as having "come in through the hawse-hole." On the other hand if he has started as a cadet, been trained as an officer, and reaches captain's rank, he is said to have "come aboard through the cabin portholes."

· GUN SALUTES ·

Were first fired as an act of good faith. In the days when it took so long to reload a gun, it was a proof of friendly intention when ships' cannons were thus dis--charged upon entering a port.

WINDAS

· HALF ~ MAST ·

Flags flown at half-mast for mourn-ing, are a survival of the old custom which decreed that slovenliness was a mark of respect for the dead. Sails and rigging were slacked off, yards cock-billed, flags part lowered etc.; in fact, anything to give the ship a dejected appearance.

· BOXING *the* COMPASS ·

Today meaning to tell off the points of the compass-card, it originally referred to the placing of the compass in a bittacle (small box). From this word "bittacle" we get our modern "binnacle."

· CAPTAIN ·

This naval officer receives his title from the old word CAPUT (meaning Chief), a name of great honor among the ancient Thanes.

• SHROUDS •

The side stays supporting a ship's masts were named "shrouds" because in early days the quality of hemp was so poor that the enormous amount of rope used for these stays actually shrouded (hid) the mast.

• HOMEWARD BOUND STITCHES •

In the old days Jack was particularly neat with his sewing. Careless sewing was referred to as "Homeward Bound" stitching, the idea being that it was only a temporary makeshift until port was reached, when mother, wife or sweetheart could be counted on to finish the job properly.

• A "SEA-GOING" INVENTION •

Canned or condensed milk was actually invented by Gail Borden in 1856, so that cows would not have to be carried aboard ships. While Borden's primary object was to produce a sufficient and sanitary food supply for babies on the voyage, the invention proved invaluable to infant and adult alike.

• BOATS VISIT ON THE • STARBOARD SIDE

The starboard side of a vessel is the traditional side to approach when making a visit by boat. This is because in ancient days ships were steered by a huge oar secured to that side, and a shipmaster, whose equally traditional station was close to the helm, could thus easily see who it was approaching his ship, and either welcome them or warn them off.

WINDAS

· ABOVE BOARD ·

This slang-term for honesty, originated in the day when pirates would sometimes hide most of their crew behind the bulwarks, in order to lure some unsuspecting victim into thinking him an honest merchantman. In reverse, therefore, anyone who displayed all his crew openly on deck, was obviously an honest seaman.

· BETWEEN *the* DEVIL *and the* DEEP ·

In wooden ships, the "Devil" was the longest seam to be caulked, and called for a bos'un's chair in order to execute the job; thus a man was actually suspended between the "devil" and the water. This slang-term was coined to describe being in an awkward situation.

· KNOCK-OFF ·

Slang for quitting work. It arose from the custom aboard slave-galleys to have a man beat time for the rowers. While he kept knocking on the block with his mallet, they rowed; when he stopped, they could cease..

· DAGO ·

A nickname which is at least 400 years old. It was bestowed by English sailors on the Portugese and Spanish seamen, who called loudly and often on their patron Saint Diego.

• PATAGONIA •

This South American country received its name when Portuguese sailors noted with amazement the huge feet of the natives, in some instances the foot equalling the length of the lower leg. They promptly nicknamed the Indians Patagones (Pata, feet) (Gones, big) and their country Patagonia.

• FOUL ANCHOR •

It seems strange that the navies of the world should use as insignia the abomination of all good sailormen: somewhere back in early days a draftsman with more artistic ability than technical knowledge produced the well-known design which shows an anchor with its cable hopelessly fouled around the shank and arms. How such a design could win the approval of the Admiralty Board is beyond comprehension, but the fact remains that the sign of the fouled anchor has become an international emblem.

• A-1 AT LLOYDS •

When you are questioned as to your health, and volunteer the cheerful information that you are feeling A-1, or A-1 at Lloyds, you are borrowing an old marine insurance phrase.

Originally, Lloyds was a little coffee shop in London, where shipowners and merchants met to discuss freights, and of course, insurance rates.

The tiny coffee house finally developed into the greatest marine insurance company in the world, which gave its rating to first class ships, by registering them A-1 at Lloyds.

Thus the term created the meaning carried in your information, that you are feeling absolutely at your best.

· SEVEN BELLS ·

American sailors are quick to note the absence of the striking of "Seven Bells" in the second dog-watch aboard British ships. In 1797 "Seven Bells" was to be the signal for the navy mutiny at the Nore. The plot was discovered and the mutiny quelled. The Admiralty decreed that "Seven Bells in the second dog-watch" should never again be struck on British vessels.

·HARNESS CASK ·

Old navy slang for the "Salt Horse" barrel. Sailors swore that horse-meat was used instead of beef, and pointed out that the extra tough pieces were parts of the horses' harness.

· TATTOO ·

Tattooing was first used as a means of identification. In the days when most poor navy men could neither read nor write their names, it meant something to be able to prove identity by means of an anchor on one's arm, or a full-rigged ship on your manly chest.

· SIDE-LIGHTS ·

It was not until between 1825 and 1830 that RED and GREEN side-lights were introduced. Up until then all ships' running lights were WHITE, but the advent of speed called for the colored lights as a further aid to navigation. . .

• STRIKE-ME-BLIND •

Old Navy nickname for rice pudding with raisins. The inference was that you were struck blind looking for the raisins.

• FIRST AMERICAN •
STEAMSHIP VOYAGE

About one hundred and thirty years ago, the first recorded sea voyage by a U. S. steamship was completed by the *Phoenix,* which actually went clear around New Jersey.

• SWEATING THE GLASS •

Sweating the glass, or flogging the clock, was an old scheme by which the sand in the hour-glass was hurried down by shaking it (or, at a later date, the hands of the chronometer were put forward) in order to shorten the time of the watch on deck. Needless to say, the scheme was not *very* popular with the navigating officer.

• BALACLAVA HELMETS •

A common stunt in Old Navy days was to cut the foot from a woolen sock, and wear the leg of it pulled over one's head in cold weather. It was called a "Balaclava Helmet" because British soldiers in the Crimean War started the idea during that freezing campaign in Russia. We believe that this was the forerunner of the modern watchcap.

· LADRONE ISLANDS ·

The Ladrones received their name from the Portugese word "Latro" (thief), and were so called because Magellan's sailors suffered so much loss of small arms and wearing apparel through pilfering by the natives.

· First NAVY BOARD ·

The first NAVY BOARD was inaugurated in the 16th. Century. Until that time, all naval strategy and fighting was directed by ARMY officers.

· The "HUSH-HUSH" FLEET ·

Nickname given the first British battle cruisers, because of the profound secrecy surrounding their building.

· SICK-BAY ·

Ship's hospitals were originally known as "Sick Berths", but as they were generally located in the rounded sterns of the old battle-wagons, their contours suggested a "bay", and the latter name was given them.

• BRIGS •

It seems a shame, but it's a fact that those handsome little two-masted square-riggers which nobly filled their quota of romance and usefulness in the days of sail, received their name "BRIG" from the word "BRIGANDINE," a term for Levant pirates or outlaws.

• CHEESE-PARING •

Here is a much used term which has lost entirely its original meaning. Today, cheese-paring describes the niggardly person who squeezes each nickel till the buffalo yells for help.

Originally, cheese-paring was petty theft. Shipmasters carrying cargoes of cheese from the Hook of Holland to London would steal the mould-ridge from the balls of cheese, and press them into extra balls for their own profits.

As they used knives in the whittling-down process, they became known as "cheese-parers," or in less polite language cheats or thieves.

• TAKEN ABACK •

One of the hazards of the days of sail has become an apt phrase in modern language to describe the feelings of a person jolted by unpleasant news. He is taken aback; his mental equilibrium is upset, and for a moment he is unable to act normally.

It is particularly apt when one considers the source of the term, for when a sailing ship was taken aback (by reason of sudden squalls or faulty steering) she was momentarily helpless and in a position of great peril. With her sails blown back against the masts, she was in grave danger of being dismasted and transformed into a derelict. Only the smartest action by skilled seamen could save her.

Yes! we repeat again, "taken aback" is one of the most apt and descriptive phrases which has been handed down from its nautical origin.

WINDAS

· WHISTLE *for the* WIND ·

In calm weather, Norsemen would whistle loudly, believing that Thor (the thunder god) would whistle in answer, thus creating a breeze which would enable the seamen to set their sail and save the arduous work of rowing. So tenacious has been this superstition, that to this day it is against regulations to whistle aboard sailing ships during a gale.

BLACK NECKERCHIEFS

Every time an American gob dons his neckerchief, he is unconsciously paying tribute to the death of Lord Nelson. This however, is only because the American uniform is patterned so closely after the British.

· SNUG HARBOR ·

A term used to describe the enviable position of a sailor, who has saved enough during his service to retire and live in comfort ashore for the rest of his life.

· FEELING BLUE ·

If you are melancholy, and describe yourself as "feeling blue", you are using a phrase coined from a custom among many old deepwater ships, by which, if the vessel lost captain or officers during a voyage, she would fly blue flags and have a blue band painted along her entire hull, when returning to home port.

• SMALL TALK...EH! WHAT? •

Major General Thomas Holcomb of the U. S. Marines can "tell it to the Marines" . . . in Chinese. His vocabulary includes some 4,500 Chinese words, almost five times the number known to the average Chinaman.

• CLAWING-OFF •

Old Navy slang for anyone stuttering and stammering, trying to side-step an embarrassing question or argument. Its significance had reference to the back-breaking task of kedging (clawing) sailing ships past headlands to catch a breeze. The task itself consisted of conveying a kedge anchor a cable's-length ahead of your becalmed vessel, dropping it over, and rowing back to the ship. Then, with the cable bent onto a windlass, you hove the ship forward to the kedge, and repeated the operation until you encountered the desired breeze. Yowsir! it was some job. *You* try it.

• SAILING CLOSE TO THE WIND •

A sinister meaning has crept into this perfectly innocent and harmless phrase.

A business man is making money by shady and suspicious methods. "Ah!" we whisper into a scandalized neighbor's ear. "He's making a fortune but—he's sailing awful close to the wind."

As a matter of fact, when a vessel is sailing close to the wind, she is merely pointing her nose as nearly into the wind as will allow headway. In seagoing parlance she is sailing close-hauled. There is nothing dangerous in so doing, nothing to write home about. Why the sinister element has been injected into the term is a mystery. It is just one of those things.

· GANGWAY ·

The gangway or gangplank received its name from the plank which extended midships from stem to stern on the old slave-galleys. Here paraded the whip-master who urged the slaves to continuous effort with triple-thonged lash.

· BELAY ·

Originally, this caution to cease hauling was "DE-lay", then followed the order "make fast". The word was later corrupted to the modern "be-lay."

· DITTY BOX ·

The ditty box (or bag) was first known as the "DITTO" bag, because of the fact that it contained TWO of everything: two spools of thread, two needles, two buttons, etc, etc.

· POUND and PINT~ERS·

Slang term for British ships when poor feeding aboard them caused Parliament to pass a law making it compulsory that every seaman be given ONE POUND of food and ONE PINT of tea or coffee at each meal.

• LIVERPOOL "PENNANTS" •

This was the nickname for the piece of string which a lazy sailorman would substitute for a missing button on his uniform. But Lord help him if he was found with such a "pennant" during inspection.

• KEEP RED TO RED •

The rule of the road at sea, to keep to the right when approaching passing vessels, dates from the time when steering oars were at the starboard quarter. Hence, when vessels approached or came alongside, the steering oar was saved the risk of damage.

• SPARKS •

Though "Sparks" is still the traditional nickname for a ship's wireless operator, modern wireless plants have removed the main reason for his being so called.

He got the name originally because of the huge spark which jumped continually between the open arcs of the early wireless sets.

• THE ICEBERG PATROL •

Because of the unceasing vigilance of the Coast Guard in this particular service, it is of interest to remember that the ICEBERG PATROL was inaugurated as the result of the *Titanic* disaster, in 1912, when that mighty vessel on her maiden voyage had her bottom ripped out by ice, and went down with the loss of over one thousand souls. From that time on our gallant coastguardsmen have scouted the North Atlantic, warning all ships of any menacing icefields.

WINDAS

• THE BOTTLE of WINE •

Christening a ship with a bottle of wine when launching her, is supposed to be a relic of the barbarous days when Norsemen were alleged to have broken the backs of prisoners across the bows of vessels being launched......this, as a peace offering to the gods.

• STERLING SILVER •

This modern term for genuine silver was originally called "Easterling Silver," and referred to a tribe on the Baltic who insisted on being paid in cash for their goods instead of by the then common system of barter. Seamen of that early date therefore carried money specially for that trade, and called it "Easterling money" or "Easterling Silver."

•BATTEN YOUR HATCH•

Old Navy slang meaning to "stop talking"; or in less polite and more modern language to "shut up."

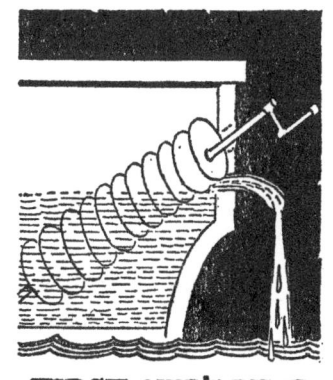

•FIRST SHIP'S PUMP•

The first known device for emptying a vessel's hold of water was invented by Archemides in the year 217 B.C.

• CANVAS •

We get the word canvas from the Latin *canabis*.

• CUMSHAW •

This is the international sea-going term for petty graft or secret commissions.

• CLIPPER SHIPS WERE • CREATED FOR GOLD

Man's cupidity, they say, has been responsible for most of his great discoveries.

This is particularly true of the beautiful clipper ships, whose graceful fabrics and time-smashing voyages placed America's merchant fleets at the peak of their prestige.

These fine tall ships were designed and built to meet the need for speed, when the world awoke one morning to find there was plenty of gold on the banks of the Sacramento.

Thousands, hundreds of thousands from every corner of the earth clamored to reach the shores of sunny California without loss of time.

Came then Donald MacKay and others of like ilk, turning their dreams into blue-prints, their blue-prints into staunch canvas-smothered craft which stormed around the Horn, and flung their gold-crazed passengers onto the waterfront of 'Frisco, days, aye! and weeks, ahead of schedule.

Cupidity may have been the primary cause for their creation, but Romance and High Adventure played no small part in the voyages of these graceful argonauts.

· The ROUND BATTLESHIP ·

Invented by an Australian in the early 70's, the plans for a battleship with a circular hull were purchased by the Russian government, and such a vessel was actually built and launched in 1876.

· First Ship on GREAT LAKES ·

The first ship to sail on the Great Lakes was a fighting ship, the GRIFFON, built and launched by La Salle in 1679

· LANYARD ·

This word was originally spelled "Land Yard, and meant a piece of small stuff of a regulation length of three feet. In other words, a landsman's measure of one yard.

·GET OUT YOUR WEB FEET·

Old Navy slang, indicating that dirty weather was ahead. It was offered as advice to young seamen by A.B's, and suggested that they should get sou'westers and oilskins out of their seachests, as rain or snow was coming.

• THE COMPASS-KEY •

"Now you run along and ask the lieutenant for the key to wind the compass. He'll be glad to see a greenhorn catching on to ship duties so smartly . . ."

• OLD GLORY'S FIRST • WORLD VOYAGE

The ship *Columbia,* commanded by Captain Gray, was the first vessel ever to carry the American flag around the world. This was when Gray circumnavigated the globe in the years 1787–1790.

• FRAPPED •

Here's another Old Navy term for an inebriate. To "frap" a vessel was to pass cables under her keel and secure the two ends of the cable on deck, tightening them *à la tourniquet* by means of a capstan-bar.

This was done when bad weather had opened big seams by starting her strakes, and "frapping" literally tied the hull together as one ties a package.

There have been some remarkable stories of voyages by "frapped" ships, the most notable being that of the vessel in which Saint Paul was travelling on his Biblical missions.

The most remarkable, however, was the voyage from Rangoon to London of an old East Indiaman, which was in such unseaworthy condition that she was "frapped" at two fathom intervals for the entire length of her hull. That she made the wild and hazardous passage down the Indian Ocean, around Capetown and back to her docks on the Thames, speaks volumes for the courage and resourcefulness of her master and his crew.

The slang term, of course, has reference to the heavy weather a drunken man makes in his progress.

WINDAS

• OIL on TROUBLED WATERS •

When this popular term (meaning an endeavor to pacify belligerents) is used, it is borrowed from the old whaler's custom (now practised generally) of putting down a coating of oil on the rough seas around a sinking vessel in order to make a safer seaway for the rescuing boats.

• MAINS'L HAUL •

This old-time Navy order to "tack ship", was also a nickname for loot or booty taken in a fight.

• NAVY or MERCHANT SERVICE •

The rivalry between navy men and merchantmen is deep rooted. As far back as Ceasar's time, when it was against the law for government ships to put to sea between the months of September–April, the navy called the merchant seamen "an avaricious lot" because they were not content to stay ashore but continued their trading despite the laws and the adverse seasons.

• ARRIVE •

The good dictionary word "Arrive" was originally an exclusively nautical expression. It was derived from the Latin word "Arripare", and meant "To come to shore."

• THE MARK OF THE BEAST •

Playful nickname for the white patch on the lapels of a British midshipman's uniform.

• BACKING AND FILLING •

This seagoing term has its shore-side equivalent in the expression "blowing hot and cold," for in our everyday speech their meanings are identical. The nautical phrase accurately expresses the motions of a sailing vessel holding her position without actually heaving to.

She would back her yards to spill the wind and stop forward motion; then, when tide or breeze tended to move her too far away, she would square her yards and run up to her original position.

So when we use the term in describing the indecision of a certain party's behavior, we are holding very well to its original meaning.

• HAND-ME-DOWNS •

Jack's wonderfully concise term for ready-made or second-hand clothing.

No other description would so completely fill the bill nor satisfy the imagination.

Visualize any waterfront town with its shoddy stores displaying garments of every description, both new and old.

Peajackets, oilskins, sou'westers, double-breasted serge coats hocked by hungry owners, or brand new sea-boots at scandalous prices. All hanging on hooks high above one's head, and all requiring the use of stepladder or chair (and a stick with a hook on it) in order to be brought down for the prospective purchaser's inspection.

Yes! "hand-me-downs" is the only possible name for merchandise thus displayed, so "hand-me-downs" it is.

WINDAS

• A FIGHTING FERRYBOAT •

When the Confederate ironclad ARKANSAS was making things miserable for the Federalists during the Civil War, Yankee genius rose to the occasion by commandeering an unfinished ferryboat and sheathing her in iron. This was the Essex, which met and sank the ARKANSAS after a hot fight.

• RIDING the BEAR •

Whenever you feel that you are overworked in Uncle Sam's modern battle-wagons, just remember that in the old Navy "riding the bear" was common practise. This consisted of filling a box-like frame with holystones and hauling it back and forth to give rough decks an extra slick-up.

•MAKING a HALF-BOARD•

In this modern era, "mechanics" have almost driven "sail" into the discard, so it would be interesting to know how many of our young seamen could "make a half-board," i.e; Run a sailboat into the wind, hold her there until all way is lost, then fill again on the same tack moving forward to your goal.

• A "HUGH WILLIAMS" •

British slang term for a sole survivor of a sea tragedy. The term is based on the fact, that, over a period of some two hundred years, nearly forty "sole survivors" have born the name Hugh Williams.

• JAMAICA DISCIPLINE •

Old Navy slang for unruly behavior. The idea of the term dates back to when pirates, returning from a successful cruise, would anchor their ships in Jamaica Harbor and waive all rules of discipline, so that their crews could indulge in a few days of unrestricted debauchery.

• PULLEY HAUL •

Let a landsman call a block a "pulley" (and he invariably does) and the nearest seaman will sneer at him for a lubber. But . . . let the steering gear carry away; let the ship be kept on her course by securing blocks and tackle direct to the rudder so that the crew can haul the rudder to port or starboard as required, and what happens? The hairiest-chested of them all will stoutly affirm they steered their ship "pulley" haul.

You figure it out. I pass.

• A "BARNEY FOKKE" PASSAGE •

This phrase means literally, "a long and protracted voyage." It gathers its inference from the fabled tale of Barnard Fokke the Flying Dutchman.

This pugnacious sea-dog was so enraged by adverse winds encountered as he attempted to round the Cape of Good Hope, that he swore with awesome oaths that Heaven itself would not turn him from his purpose.

Heaven countered (or so the story goes) by decreeing that Fokke should keep on trying to weather the Cape until the Day of Judgment, and from all accounts of his ghostly ship and ghostly crew as recorded by honest mortal seamen from that day to this, it would seem that the poor swab is still trying.

WINDAS

• FINANCING COLUMBUS •

Contrary to popular belief and most historians, Queen Isabella of Spain did not pawn her jewels nor advance one penny to finance the discovery of the New World. Luis de Santangel, keeper of the Ecclesiastic Revenues of Spain loaned Columbus the money for his famous voyage. Incidentally, it was fortunate for Santangel that Columbus was successful, otherwise little Luis would have been executed for misuse of the funds in his keeping.

• KNOW YOUR ENSIGN •

Did you know that for 23 years, Old Glory carried 15 stripes? This was from 1795 until 1818, when the original 13 stripes were restored permanently.

•UNLUCKY NUMBER•

To those who claim that the number 13 is unlucky, it seems more than mere coincidence that the shell which blew up His Majesty's ship HOOD, traveled 13 miles.

• NOT SO HOT •

Germany "didn't do so good" in her last "shootin' war" with us. The only shell to strike U.S. soil was fired from a U-boat and landed on Cape Cod, Mass., in 1917.

• OIL FOR THE SIDELIGHTS •

I guess it happened to you too, during your first voyage . . . being sent to ask for RED and GREEN oil for the sidelights. Or was *I* the only sucker?

• THE STORM KING •

In 1850, Sir William Reid of England formulated the "Law of Hurricanes" by comparing the logs of hundreds of sailing ships and correlating the data found therein.

• CHINESE COMPASS •

Just in case you think the compass is a modern invention, remember that the Chinese used a compass of magnetized iron 200 years B.C.

• MARINES TAKE NOTE •

The first U. S. Marine Corps was comprised of hard-hitting hairy-chested men of the merchant service.

To John Paul Jones is credited the idea of recruiting seamen from trading and fishing vessels to fill the ranks of a sorely needed company of sea-soldiers. From all accounts they were hard to train and tough to handle. They scoffed at discipline, jeered at the idea of saluting, and thumbed their noses at martinets. They handled their muskets as though the weapons were capstan bars, and their parade formations were a nightmare to the drill instructor. But they proved their mettle in many a blistering sea-battle, and left a heritage of loyalty and courage which still burns brightly as a beacon to guide our modern "Devil Dogs."

WINDAS

• FIRECRACKERS •

Next time you are celebrating the Fourth of July, remember the days of your youth, and salute the memory of the United States ship GRAND TURK. This famous vessel was not only the first American ship to visit China.... she was the first ever to bring home a load of firecrackers, thus fostering our tradition for pyro-technic display.

•PENNY ROYAL•

This famous oil, universally recognised as an anti-dote for mosquito bite, was mis-named by the illiterate seamen who discovered its beneficial use in Havana in early days. It is now known only as Penny Royal, but its real name is PENNER OIL.

• IRON SHIPS •

Remember, when you are slipping through the water in your mighty steel battlewagon, that only 12 decades ago seamen protested to Parliament against sending iron ships to sea on the grounds that "Everyone knows that metal cannot float."

ANNIE DOESN'T LIVE HERE ANYMORE!

The famous old school-ship "Annapolis (known to hundreds who served their cadet-ship in her as "Annie") has been re-tired and a steamship substi-tuted in her place. Built in 1896 the "Annapolis" served in the Spanish-American War and also in the first World War

• FIRST SHIP TO FLY THE •
U. S. ENSIGN

The first ship to fly the Stars and Stripes was the *Ranger*, when she was commanded by the redoubtable John Paul Jones.

• GREENWICH TIME •

Although Greenwich was established in 1675 as the international time center for mariners, it was not until the first World War, 1914-1918, that French navigators used any other than Paris time for their reckonings.

• COCKBILLED •

Old Navy slang describing drunkenness. It had reference to the old custom of cockbilling (hauling askew) a ship's yards to make room for swinging cargo or supplies aboard. The inference of the term, of course, was directed at the slantwise posture of a man walking while under the influence of liquor.

• HE WENT HOME D.B.S. •

Nowadays this expression signifies generally the arrival home, flat broke, of any seaman.

Its meaning lies in the legally required custom of British shipowners, that when any of their vessels are wrecked they must pay for the crew to be brought back to the home port.

The sailors are returned under the Distressed British Seamen's Act.

GEORGE WASHINGTON
almost A SAILORMAN ·

"The Father of his Country" was a land-lubber through no fault of his own. As a young man he was so determined to join the navy (in spite of parental objections) that he allowed himself to be taken in a waterfront saloon by the press-gang. However, he was doomed to disappointment, for the officer in charge of the press-gang recognised him as the son of an influential citizen and ordered him to go home.... much to George's chagrin.

· COFFIN SHIPS ·

This term is in no way descriptive of vessels loaded with wooden kimonas. It is a maritime insurance designation and refers to the corrupt custom of over-insuring un-seaworthy hulls, and sending them on a voyage with the sinister purpose of deliberately losing them.

·FIRST NAVAL BATTLE·

The first historic naval battle was fought in 480 B.C., when 366 Greek ships outfought and overpowered 600 Persian vessels at the Battle of Salamis.

· ENSIGNS ·

Though today we look on ensign as a purely naval rank, as a matter of fact an ensign was originally the lowest commissioned officer in the Army.

• MEDITERRANEAN CHARTS •

The first known charts to be used by European seamen were made by the Italians in 1351, and were known as *Portolani*.

• DAVEY JONES •

This name is a corruption of Jonah, the Biblical gentleman who is credited with having supplied Vitamin A to the whale for three days.

• THE FIFE-RAIL •

The pin-rail at the base of a mast was named for the little fife-and-drum boys who perched there out of the way of their elders, while men-o'-warsmen maneuvered or drilled to the shrill note of the fife and drum tattoo.

Fife and drum have long since disappeared, but the rail retains the name of the boy who sat and played so bravely.

• CATAMARAN •

This small vessel, used by the natives of Ceylon and adjacent areas, is one of the most remarkable surf boats known to seamen.

It is formed of three logs of lumber, secured together by three spreaders and cross lashings.

It is about twenty feet long as a rule, and two men can easily handle it.

Frail as the structure may appear, it can ride the heaviest surf at Madras where even a staunch life boat might be apt to capsize.

WINDAS

A "CHEESEY" SEA-FIGHT

In 1841, when warships of Uruguay and Argentina battled for sea supremacy, the Uruguayan fleet, in one action, ran out of ammunition. Undaunted, they substituted hard round cheeses for cannon-balls and actually came out victorious in the engagement.

SHOW A LEG !

This slang term for ordering men to turn out, originated in King George Third's time, when women were allowed to accompany sailors on long voyages. It was customary when ordering seamen from their bunks, for the bos'un to demand "Show a leg." If the leg was covered by a stocking he knew it belonged to a woman; otherwise the skulker would promptly be routed from his bunk.

AMIDSHIPS

Though there is a common tendency these days to misname the waist of a ship "midships," all good sailormen should know that "midships" is an imaginary deckline running parallel with the keel from stem to stern. Thus, the masts are stepped "amidships."

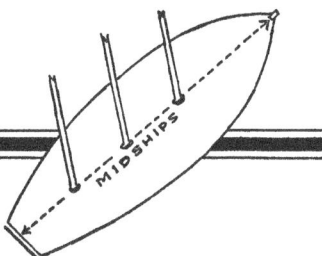

MIDSHIPS

SQUARED AWAY

When you use this expression for putting yourself into a posture of offense or defense, you are borrowing a phrase which described a square-rigged ship bracing her yards to run away before the wind.

• POOP-DECK •

A ship's after-deck receives its name from the old Roman custom of carrying *pupi* (small images of their gods) in the stern of their ships for luck.

• THE BEST GIRL'S ON • THE TOWROPE

Old Navy slang descriptive of "fair weather and easy going," on the homeward passage. If the ship was making good progress, the seamen would say "Ah! the best girl is on the towrope."

• SAILING UNDER FALSE COLORS •

In view of the acts of enemy raiders flying neutral ensigns, it is of interest to note that this was first the subterfuge of pirates. These seagoing gangsters would fly the flag of some friendly nation in order to lull prospective victims into a feeling of security.

• SWALLOWED THE ANCHOR •

For the benefit of non-seagoing readers, let us state that when a seaman uses this expression, he is not trying to belittle sword-swallowers or others of similar professions by claiming to have actually swallowed the right or left bower.

A sailor has "swallowed the anchor" when he has permanently quit the sea to take up a shore job, or to retire.

WINDAS

SPINNING A YARN

This term for tale-telling, was coined in days when sailors would be given old ropes to untove for the making of sennit and small stuff. As this was the only duty during which they could talk at will, the act of making yarn became synonomous for free and unrestricted conversation.

MATE

Bosun's Mate, Gunner's Mate, Mate of a ship, all derive their rating from the French word "Matelot" meaning sailor.

THREE COLLAR BRAIDS

The three white braids on the American sailor's collar really commemorate three big British naval victories, viz:- Battle of the Baltic, Battle of the Nile and Battle of Trafalgar; and this is because the American uniform is practically a duplicate of the British.

FIRST SHIP'S WHEEL

While history fails to record the exact date when wheels took the place of tillers for steering, it is generally conceded as somewhere between 1703-1747

• CHIPPING WITH A RUBBER • HAMMER

This was an old-time superstition regarding eternal punishments for wicked seamen, and which pictured the souls of the evil ones forever hopelessly chipping . . . with a hammer made of rubber.

• FIRST TELESCOPE •

The first telescope was invented by a Dutch spectacle maker. Experimenting in 1608 with one lens placed over another, he accidentally discovered the optical principle for bringing distant objects closer to one's vision.

• DANDYFUNK •

Any time you feel disposed to growl at the dessert set before you, just remember that "Dandyfunk" (a messy concoction of broken ship-biscuit smeared with molasses) was practically the only dessert ever served aboard the old wooden ships. Ask gran'pappy, he knows.

• THE "HORSE" LATITUDES •

Roughly speaking, that area between North Latitudes thirty and forty degrees. The term was coined in the days when sailing ships carrying horses from Europe to America were often becalmed in the region of the West Indies. To conserve drinking water for the horses, some of the animals would be thrown overboard in an effort to save the others.

PLUM DUFF

Some bright sea-cook once decreed that if R-O-U-G-H spelled "ruff," and T-O-U-G-H spelled "tuff," then D-O-U-G-H must spell "duff." Thus ships' dessert received its traditional name.

CAPE HORN "FEVER"

Old navy slang for malingering when in the latitude of "Cape Stiff." Sailors developed the habit of acquiring sudden and mysterious maladies in order to avoid working on deck in these bitter regions. Needless to say, officers discouraged the habit.

·BO'SUN· In the 17th. century, ships were required by law to carry THREE boats, and named respectively ① the BOAT ② the COCK ③ the SKIFF. The men in charge rated BOATSWAIN COCKSWAIN and SKIFFSWAIN. Swain meant lover or keeper.

·THE PAINTER· The boat painter receives its name from the French word "Peyntours" meaning noose or bight.

• BOMB ALLEY •

British Navy slang for the Strait of Sicily; bearing in mind the almost continuous sky-attacks on British shipping in this area in 1941, the inference is obvious.

• OVERWHELMED •

This word, meaning crushed or defeated, is from the old Anglo-Saxon *whelman*, which means to turn a vessel completely over.

• NOT ROOM TO SWING A CAT •

This phrase, today descriptive of any very cramped quarters, was coined in an era when it was often customary to flog men in the ship's brig. If the brig was too small to allow the sergeant-at-arms full play for his cat-o'-nine-tails, the culprit would be taken on deck and there punished for his misdeeds.

• GROG ON SWEDISH SHIPS •

Trusting that this information will not result in wholesale desertion from Uncle Sam's ships, we mention with some hesitance an old custom aboard Swedish vessels; that of having a bucket of grog placed at the wheel for whoever wants a drink, between the festive days of December the twenty-fourth and January the first.

The THREE MILE LIMIT •

In view of the current controversy over the gambling ships, it is of special interest to note that the reason for three miles being the distance over which a nation has jurisdiction regarding coastal waters, is because at the time this international law was established, three miles was the longest range of any nation's largest guns, and therefore the limit to which they could enforce their laws.

• COMMODORE •

This title arose from a practised economy of the old Dutch Admiralty. In her war with England, Holland found herself short of admirals and distressingly short of cash. She solved her difficulty by creating a brand new rank of "Commodore", which carried with it all the responsibilities of an admiral ___ but only HALF his pay.

• ANCHOR •

Anchors received their name from the ancient Greek word meaning "crook or hook", and old Grecian anchors were actually in this crude form.

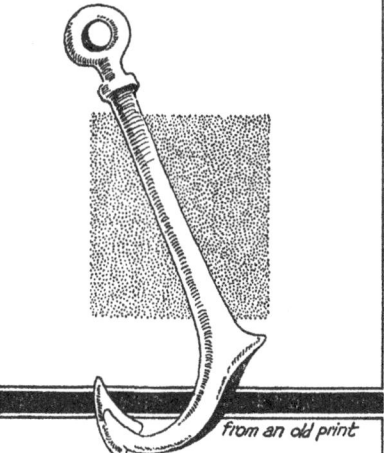

from an old print

• 'SHE-OAK' NET •

A slang term from "Down Under," for the safety-net slung under gang-planks. The name is derived from a brand of potent Australian beer, which carried a picture of a she-oak tree on the bottle label. As the beer had an alcoholic content of some 15 or 20 per cent, one can very easily understand why the safety net was so nicknamed by "Aussie" sailors returning from a particularly hilarious shore leave.

• OLDEST SHIP AFLOAT •

The oldest ship afloat is a vessel which was actually built in 1580 and presented to Ivan the Terrible by Queen Elizabeth. It is still known as "The Little Father of the Russian Fleet."

• CHIEF EXPORT . . . • MASTER MARINERS

Sandy Cove, Nova Scotia, does a thriving business in sea captains. With an estimated population of some two hundred people, Sandy Cove has sent to sea fifty-five master mariners.

If that's not over-subscribing one's quota, we'd like to know what to call it.

• PROVIDER . . . ON A BIG SCALE •

In the thirteenth century Irish shipmasters were compelled by law to provide for the needs of the families of each member of their crews. We don't know if this included "his sisters and his cousins and his aunts" but it was a tall order anyway.

• SPLICE THE MAINBRACE •

During old sea battles, a vessel's rigging was a favored target, and the first duty following an engagement was to set up broken gear and repair sheets and braces. It was the custom, after the main braces were spliced, to serve grog to the entire crew. Today the meaning of this old custom has been twisted into a general invitation to "have a drink," or as the saying goes "splice the main brace."

WINDAS

The CAPE of GOOD HOPE

Because U.S. seamen may be seeing a lot of the Cape in the near future, it is of interest to note that originally this promintory was known as the Cape of Torments. Thus it was called by Bartholomew Díaz who first braved its bitter dangers in adverse weather. But his king changed its name to that of Good Hope, saying "its discovery promises great and lively hopes of wealthy lands to add to the realms of Portugal."

The SUN'S over the FOREYARD

Old Navy slang for "It's time to have a drink." It derived its meaning in the days when drunkeness was common aboard ship, so the Admiralty ordered "no officer shall partake of liquor until the sun shall have risen well above the foreyard".

CONNING TOWER

The control center got its name from a corruption of "Cunning", and literally referred to the cunning or cleverness of a sailing master in maneuvering his vessel.

FIRST AMERICAN ENSIGN

The flag first raised by John Paul Jones was originally designed for the private use of the "Honorable John Company" (British East India Co.)

• AMMUNITION •

This common word for gun charges is derived from the French *munire*, meaning "to provide."

• SWISS NAVY •

Though it is a traditional joke among sailors that "I served my first hitch in the Navy of Switzerland," such a navy actually existed in 1799, when an English Captain Williams commanded a fleet of small vessels on Lake Zurich, in operations against the French.

• "PETTICOAT" POINT •

Way back in 1915, this might have been a good nickname for Navy Point, N. Y. . . . but don't ever say *we* said so. In that year Mrs. Metcalf took over command of this navy yard on the death of her husband, and thereby created a precedent, as no other navy yard had ever been commanded by a woman.

• LOADED TO THE GUARDS •

Slang term for a man who has imbibed as much liquor as he can conveniently carry.

It has reference to a ship's load line or Plimsoll marks. These are painted clearly on every ship's hull and indicate the point at which the vessel must be guarded from over-loading.

With safety, she can carry just so much, and no more.

It's an apt term, sure enough.

WINDA B

The WOMEN of FIDDLER'S GREEN

Just in case you didn't know it, calm weather and smooth seas are the result of the sweet songs of the women of Fiddler's Green, who sing to keep the waves in unison. Whenever they stop singing, the waves get restless, and bad weather ensues. Next time you're enjoying a pleasant voyage, remember that Davey Jones' glamour girls are really doing their stuff.....for which be thankful.

· SKYLARK ·

This perfectly good dictionary word was originally a slang term, used to describe the antics of lusty young navymen, who would slide down the backstays for fun. The latter half of the word is from the ancient "LAC", meaning "to play".

· STEAMSHIP ·

Strangely enough, the inventor of the steamship was neither a sailor nor an engineer. Robert Fulton was a nationally recognised ARTIST, who forsook Art to devote his life to his wonderful invention.

· SCHOONER on the ROCKS ·

In old Navy parlance, a "Schooner on the Rocks" was a roast of beef surrounded by baked potatoes.

• HOLY JOE •

This traditional shipboard slang for a preacher is generally supposed to have been coined for Joseph Smith, founder and Prophet of the Mormons.

• THE "BURNT OFFERING" •

Limey slang for any plain roast of beef or mutton. Judging from personal experience with some of those old British sea-cooks we'd say the nickname registered A-1.

• HAWSE HOLES •

Hawse holes were originally eyes carved or painted on the bows of ancient ships for the purpose of allowing the vessel to see if there were evil spirits ahead, and to veer away from her course until they had been left astern.

• STEADFAST AS A ROCK •

Talk about bulldog tenacity . . . Barbados is the only territory in the Caribbean that has remained faithful to one nation for over two hundred years.

It still flies the Union Jack, as it did when it was claimed for England by a British admiral in the seventeenth century.

All other lands in the area have switched fealty from time to time as Spain rose and fell, as the banners of Holland and Portugal held sway, or traded their holdings with the Tri-Color of France.

Only Barbados remains unchanged.

· CRAFT TO INVADE ENGLAND ·

The barges which were to carry Napoleon's troops across the English Channel when the Corsican planned to invade England, were powered by windmills, which in turn drove huge side paddles. However, they were never used, for Napoleon, like Hitler, got cold feet.

· A "SUNDOWNER".

Nickname for those tough Old Navy martinets who insisted that the entire crew be aboard "by sun-down prior to the day we sail."

·BRAZIL·

This South American republic was named for a wood found by early navigators. The wood was known as "BRAZA", meaning a live coal.

·FIRST LIQUID COMPASS·

The first liquid compass was invented and used by Arabs in the year of Our Lord 1242.

• WATERSPOUTS •

Just in case you didn't already know it, any Chinese sailorman can tell you that waterspouts are caused by naughty nautical dragons leaping up out of the sea, in an evil attempt to pull good spirits down from heaven.

• FIRST SHIP THROUGH THE • PANAMA CANAL

Just to keep the records straight . . . the first vessel to pass through the Panama Canal was the crane ship *Alex Lavalley* on January 7th, 1914.

• PROGRESS IN STEAM •

Here is an interesting and concise tabulation of the progress made by steamships since their invention in 1788.

It is now definitely established that the first steamship was tried on the Dalswinton Loch, Scotland, in 1788.

Next came the *Charlotte Dundas,* built and launched by Symington in 1802.

Then in 1806 Robert Fulton operated his American-built *Clermont.*

In 1812 the *Comet* went into regular service on the river Clyde.

In 1821 a steamship crossed the Atlantic for the first time.

The *Enterprise,* in 1838, steamed almost all the way from England to Calcutta.

Finally the *Sirius,* in 1838, flung the last remnant of canvas into history, when she made her remarkable voyage from Liverpool to New York under the sole driving power of her engines.

For the next hundred years, progress in the size of steam-driven vessels was rapidly increased from the 700 tons burthen of the *Sirius* to the mighty 54,000 tons of the *Queen Mary.*

· PRESS-GANG "PETS" ·

In old press-gang days, seamen engaged in dockyards or canvas-lofts were given papers exempting them from "pressed" service. As press-gangs were naturally anxious to get ex-seamen rather than landsmen with no experience, a press-gang "PET" was anybody in the "exempt" classification who forgot to carry the necessary papers on his person, for then, despite his protests, he could be carried off to serve.

· HORN-PIPE ·

This sailor's dance was named for the two instruments which constituted about <u>all</u> of the orchestra usually found aboard old deepwatermen.

· STEEL MASTS and RIGGING ·

The first ship to have steel masts and rigging was the SEAFORTH, built in 1863.

· SCRIMSHAW ·

How many modern seamen go in for Scrimshaw, i.e. the art of carving models etc., from bone or other material? In old days a sailorman who wasn't handy with his knife wasn't worth his salt. Just out of curiosity we'd like to get some answers in reply to the above question.

• FIRST STEEL SHIP •

The first steel ship of note was the *Servia,* launched in 1881.

• EVOLUTION OF PROPELLERS •

In 1843 the single screw was put into use as the driving agent for steamships.

In 1888 twin screws were adopted.

Then, in 1905 and 1906, respectively, triple and quadruple screws were installed.

• A "QUARTER-DECK" VOICE •

Contrary to popular belief, a "quarter-deck" voice does not refer to the stentorian volume of a man's vocal cords.

It is quite an old expression and was coined to describe the voice of authority. The term carried also the suggestion of the cultured or educated voice of an officer, as compared with the uncouth tones of an ordinary seaman of that day.

THE REGULATION 15 PER CENT

To those who do not already know, it will be of interest to note that there is a regulation "ceiling" on profits in certain departments aboard our fighting craft.

Tailor shop, laundry, barber shop, soda fountain, etc., must operate on a basis of no more than 15 per cent.

In turn this profit goes for the purchase and maintenance of athletic equipment, phonographs, and radios. It is also used to finance parties, dances, and other social functions.

WINDAS

· FLOATING MINES ·

These implements of naval warfare are by no means new. As far back as 1583 at the siege of Antwerp, boats were filled with explosives and floated against the enemies ships. They were given the very appropriate name of INFERNALS.

Can't you alter your handwriting a little bit, Bill? Those signatures look awful SIMILAR

· a "WIDOW'S MAN" ·

A form of graft very popular in the early part of last century, when imaginary sailor's names were used to "pad the payroll" at unscrupulous navy hospitals.

PUBLIC REMINDER Nº 1 (for BOOTS)

Remember! You serve IN a ship... not ON her!

· "TURNPIKE SAILOR" ·

Slang term for beggars who bum a hand-out on the false assertion that they are old seamen in distress.

• FIRST IRON SHIP •

The first vessel whose hull was built of iron was the Cunard Liner *Persia*. This was in 1855.

• FIRST SCREW-PROPELLED SHIP •

The screw as a propelling agent was first used in the *Great Britain* in 1843.

• HOUSEBUILDERS TURNED • SHIPS' CARPENTERS

When, recently, the Pennsylvania Historical Commission finished restoring the hull, decks, and interior of the famous fighting ship *Niagara*, an interesting sidelight on her construction was revealed.

The *Niagara*, flagship of Admiral Perry in the Battle of Lake Erie, was laid down from the plans of Henry Eckford, naval architect. At the same time nine other keels were laid. The government at Washington, however, failed in its promise to Captain Daniel Dobbins to send skilled shipbuilders to rush construction of this fleet on the shores of Lake Erie.

To Eckford and Dobbins, each day wasted was a nightmare. Finally the impulsive captain swore he would wait no longer on the government and, taking horse, rode around the countryside rounding up all the house carpenters he could find.

It was a makeshift arrangement, but enthusiasm and patriotism outweighed any lack of skill on the part of the workers.

The *Niagara* and her sister ships were built, and well built, by men who had never previously constructed a ship's hull in their lives; and they were completed in time for Admiral Perry to gain his notable victory on Lake Erie in 1813.

· AURORA BOREALIS ·

Just in case you didn't know it, the Aurora Borealis, or Northern Lights, are the reflection of the engine-room fires from all the steam-propelled ships ever sunk. Their furnaces are kept going by Admirals, Captains, and all others of the useless Deck-gang, while worthy members of the Black-gang (engineers, tenders, et al.) loll around doing no work whatever, enjoying themselves with champagne, music and beauteous females.

·LOBSCOUSE ·

If you're inclined to grumble at the menu aboard your battle-wagon, just recollect that "dad" was fed "lobscouse" as a regular diet. It was a concoction of broken ship-biscuit, chopped up potatoes and left-over meat ends.

· JACOB'S LADDER ·

This name for a boat ladder is borrowed from the dream of the famous Biblical character, Jacob. In his dream, he saw a ladder ascending from earth to heaven, and because most of us can remember what an awful long climb it seemed, our first time aboard ship via this route, one can readily appreciate the significance of the nickname.

· TONNAGE ·

Originally this word had nothing to do with the weight or displacement of a vessel. It merely denoted the size of a ship by the number of TUNS (barrels) of wine which she could stow in her hold.

• LEARNING THE HARD WAY •

It seems strange how often the human race must be jolted by severe shocks into looking after its own welfare.

Despite centuries of sea-hazards, it took the colossal tragedy of the *Titanic* in 1912 to bring about the very obvious regulation that vessels be required by law to carry sufficient life-boats and rafts to accommodate their passengers and crew.

• SIDE BOYS •

In "tending side" duties, it would be well for our lads to remember that originally side boys had one specific duty.

Some of the officers, particularly those of higher rank, would attain considerable, shall we say . . . *embonpoint* in their later years. This made coming aboard a particularly strenuous exercise, so the side boys had the job of hauling the short-o'-breath gentlemen inboard.

• SIZE AND WEIGHT OF • RUDDERS

To the landsman, as to the apostle Saint James, it must appear astounding that so comparatively small an object as a rudder should control the movements of a huge hull.

Nearly nineteen hundred years ago, Saint James voiced his surprise when he wrote, "Behold also the ships, which though they be so great, yet are they turned about with a very small helm."

All of which is very true, and yet actually a modern liner's rudder is enormous, measuring approximately eighty feet high and fifteen feet wide.

It weighs well over one hundred tons, almost half the total weight of the *Santa Maria,* flagship of the intrepid Christopher Columbus.

• CLOSE QUARTERS •

This term, today indicative of hand to hand fighting, was originally "CLOSED Quarters" and referred specifically to special deck-houses to which the crew could retire if boarded by superior numbers. The doors were barred and loopholed, so that a deadly fire could be poured into the enemy.

• MOORINGS •

A contribution to our nautical vocabulary from the Netherlands. It is from the Dutch "Marren", meaning to tie.

TAKING A SIGHT

Old Navy slang for thumbing one's nose (discreetly and behind his back, of course) at an officer.

AMAZON RIVER

The Amazon River got its name when Admiral Orellana reported his crew attacked there by the huge native women. These female warriors were never heard of before (nor since) but the river's name stands pat.

• GUN SALUTES •

Gun salutes are fired in odd numbers . . . 1, 3, 5, 7, etc., because of the old superstition that uneven numbers are lucky.

• "DOCTOR" DISCHARGE •

Slang term for a fake discharge. In the last days of sail, when experienced sailors were at a premium, ordinary seamen would pay the ship's cook to alter their discharges into the higher rating of A.B.

As the cook is traditionally called "Doc" aboard ship, the faked tickets became known as "Doctor" discharges.

• HARD TO TAKE •

There's a little island in the Mediterranean, sixty miles off Sicily, whose glory will never fade.

It is seventeen miles long by eight miles wide, and every foot of it is battlescarred and bloody.

Since it fell into the hands of the Phoenicians in the sixth century B.C., it has resounded to the clash of arms, as Carthaginians, Romans, Arabs, Normans, Italians, Spaniards, and Frenchmen each in turn conquered and played their little part in the caves and shelters of its rocky terrain. Finally the English took over, and from that day Malta has been a bulldog guarding faithfully the interests of the Crown in the western Mediterranean.

The Germans found Malta easy to bite into but too tough to swallow.

From 1939 to the end of '42, they bombed the island consistently in more than three thousand air raids. But the only result was a net loss to the Nazis of better than one thousand planes, which was no great source of satisfaction to Mister Hitler.

The bulldog won't let go!

· WALLOP ·

When the French burned the town of Brighton (England), King Henry the 8th. sent Admiral Wallop with a swift English fleet to ravage the French coasts in reprisal. The old sea-dog made such a complete job of it, that to this day any devastating blow from an enemy is referred to as "an awful wallop."

· FUTTOCK SHROUDS ·

The name of the short shrouds extending below (and securing) the lower edges of the tops to the masts, is a corruption of the word "foot-hook."

·MOTHER CAREY·

The name of the good angel who protects Jack at sea, is an English corruption of the Portugese "Mata Cara" (Dear Virgin).

· SCHOONER ·

Today referring exclusively to vessels of fore-and-aft rig, originally meant nothing of rig, but denoted a shallow draught vessel which literally "schooned" (skipped) over the water.

• PADDY WESTER •

Here is an old-time term of opprobrium for a seaman whose lubberly behavior belied the rating recorded on his discharge.

It came into use in the '80's, and had reference to a scoundrelly boarding house keeper in Liverpool named Paddy West.

Sail was on the decline; seamen for deepwater square-riggers hard to get. Masters waiting to clear would pay big money to any crimp who could provide a crew. Paddy had a bright idea. He traversed the slums of Liverpool and gathered up the outcasts of that city on the Mersey. Drunks, barflies, the importunate, and the needy . . . took 'em all to his boarding house and fed 'em. But he was careful to lock them up at night, so that none might escape back to their natural habitat, the gutters.

In his dining room was a table above which was suspended a pair of varnished cow-horns.

Each outcast was made to walk around the table so that when the time came he could truthfully say he had "been around the Horn."

In an enclosure back of the house, Paddy set up an old ship's wheel. Here the outcasts stood and swung the wheel until they had digested the rudiments of "port your helm," "up helm," etc., etc.

Paddy's wife now enters the picture. This old harridan would perch on the fence near the embryo helmsman, and throw buckets of salt water over him, to give his pitiful clothes the appearance of being caked with brine.

The training period was now finished. The dead-beats were now fully qualified A.B.'s, and received a cleverly faked discharge which testified to same.

Remained then only for Paddy to contact the masters and mates of the waiting ships.

Sure! He had plenty of men. All A.B.'s and crackerjack helmsmen. Plenty of 'em been "round the Horn" half a dozen times.

"How much? Well, now, 'Honest Paddy' they calls me, and I wouldn't want ter take advantage of the good captain's needs. Shall we say three month's advance on each man's wages? . . ."

· U.S.S. ALFRED ·

Strangely enough, the flagship of the Revolutionary Navy was named in honor of King Alfred, and yet was commissioned to fight against the British Navy which was originally founded by that old-time English monarch.

· MAGNETIC COMPASS ·

The Chinese also invented the magetic compass some sixty years B.C., by floating a magnetized needle on a piece of cork in a bowl of water.

·ANCHORS·

The anchors invented by the Chinese nearly 4,000 years ago, were remarkably similar to the more modern Admiralty anchor.

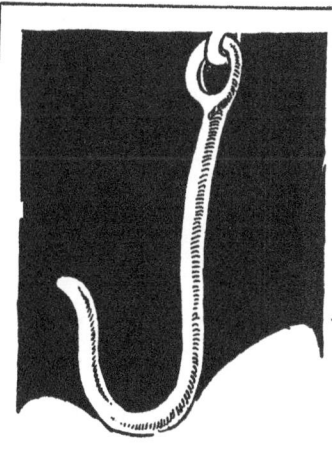

· GADGETS ·

This well known word was originally the nautical name for hooks, and derives from the French "Gâche".

• DUNNAGE •

Nickname for a seaman's personal gear, including his clothes.

The term derives from the lumber known as dunnage, used in shoring up and dressing the cargo stowed in the ship's hold.

• SLING IT OVER •

Colloquialism for "pass it to me." To seamen forever watching the loading or unloading of the ship's cargo by means of a net or sling, the latter was the obvious word to embody in a simple request like "sling me the salt."

• THE TIE THAT BINDS •

This expression of sentiment, regarding blood relationship or a similarity of ideals which hold people in a common bond, is generally believed to have been coined for the short chain which secures main and fore yards to their respective masts.

• A "SCURVY" TRICK •

Here is a commonly used phrase with a distinctly nautical origin.

In the old days of sail and interminable voyages, lack of fresh vegetables and clean drinking water made scurvy one of the most dreaded diseases which beset the deepwatermen.

It was a loathsome plague, so that the word scurvy is certainly apt, when the phrase is used to describe a particularly dirty deal that has been perpetrated.

· CROW'S NEST ·

The ship's lookout station was named for the cage which housed the ravens carried by Norsemen at their mastheads. When these sea-warriors lost sight of land, they would release one of the birds, and as it headed for the nearest shore, they would follow its flight. A crude method of navigation, but, within limits, both efficient and practical.

· HARRIET LANE ·

Slang term on British ships for canned mutton. The name refers to an Australian victim of a particularly brutal homicide, who was chopped into small pieces by her murderer. British sailors swear that the pieces were then ___ ___ but we won't go into that.

·AMERICAN INDIAN·

The original American race received its name because of a great sailorman's mistake. When Colombus discovered America, he actually thought he had won through to India, and naturally enough he named the natives "Indians."

·HE KNOWS THE ROPES·

Today a phrase indicating that a man is expert, it originally meant exactly the reverse. In very early days, when "He knows the ropes" was written on a seaman's discharge, it meant that he was only a novice, but knew the names and uses of the principal ropes.

• THE GROWTH OF WARSHIPS •

The *Royal Harry,* built in Henry the Eighth's reign, was the first English warship to attain a displacement of one thousand tons.

• BOWSPRIT •

The bowsprit, to which sailing vessels secured the figurehead, derives its name from the old Saxon world "Sprit," meaning "To sprout."

• SHORE CLOTHES •

Officers from U. S. merchant ships invariably wear "civvies" when going ashore. This follows the tradition of early days when it was done to fool "press-gangs" who were anxious to seize experienced-looking seamen to man warships.

• TWO OF A KIND •

Named for a hero of the War of 1812, the first United States destroyer *Shaw* was cut in two in 1918, but successfully navigated her way back to port, sailing the last forty miles of the voyage stern first.

The second USS *Shaw* was a victim of the treacherous Japanese attack at Pearl Harbor on December 7th, 1941.

With almost half her hull torn asunder by bombs, she gave the lie to the Nipponese report that she had been sunk by sailing halfway across the Pacific with what might better be described as a "jury" bow.

She was repaired and left port to fight again, so it appears that her sinking, like Mark Twain's death, was grossly exaggerated.

WINDAS

· HAMMOCKS ·

Swinging beds for sailors were first used by Colombus, who discovered their practical use from natives in the West Indies.

· HOLYSTONE ·

The sandstone formerly used for scouring ships' decks, got its nickname from some witty sailor who declared that as its use always brought a man to his knees, it sure must be "HOLY"

· GROG ·

Admiral Vernon wore his cloak of groggam (mixed silk and wool) so habitually, that his men nicknamed him "Old Grog." When this officer suggested that the Government could save money by diluting the Navy rum-ration with 50% water, and a law was passed to that effect, sailors, jeering, called the new ration "GROG," and the name stuck to this day.

· BUTTONS on SLEEVES ·

Midshipmens' sleeve buttons are sewed on thwartwise because in Admiral Nelson's time midshipmen had no pockets in their uniforms and therefore no place to carry handkerchiefs. To discourage the bad habit of wiping noses on sleeves, Nelson ordered buttons sewed in such position that offending noses would be hurt. ·

• BOTANY BAY •

Botany Bay, five miles south of Sydney, Australia, was named by the famous navigator, Captain Cook, because of the great number of strange new plants found there.

• MASTER MARINERS •

There were no licensed masters in charge of ships until the year 1450, when Charles the Fifth of Spain signed a law making it compulsory for a shipmaster to carry a certificate recording his qualifications for the job.

• FOR THE FOLKS BACK HOME •

All down through the ages, seamen of every nation have gathered trinkets in strange ports as presents for the loved ones at home.

Something that dad would treasure . . . or a bit of jewelry for mother and sister.

Junk, most of it, and of no intrinsic value; but very precious junk to the receiver, as a pleasant memory of the lad who brought it.

Digging into the rubbish pile of a long-vanished Indian tribe at Drake's Bay, forty miles north of the Golden Gate, University of California scientists have found hand-forged iron spikes and Chinese pottery . . . last traces of a shipwrecked Spanish galleon.

The fragments were identified, after almost a year's research, as coming from the galleon *San Augustin,* which foundered off the coast of California in November, 1595, after a tempestuous passage from the Philippines.

Maybe the spikes were for dad's new barn on the far-off hills of Andalusia; or the Chinese pottery was to have graced the walls of mother's little garden in Madrid . . . *quien sabe?*

· HUNKY~DORY ·

This term, meaning everything is O.K., was coined from a street named Honki-Dori in Yokohama. As the inhabitants of this street catered to the pleasures of sailors, one can readily understand why the street's name became synonymous for anything that is enjoyable or satisfactory.

· NORMANDY ·

This French province was named for the fierce sea-raiders who drove south along its coasts, and having conquered it, settled there; namely the Norsemen.

·RUDDERS at STERN·

Until the 13th. century, rudders were always affixed at the starboard quarter. In 1242 however, the ship "City of Elbing" was fitted with a rudder secured at her stern.

· DECKS PAINTED RED ·

In John Paul Jones' day, decks, bulwarks and gun-ports were all painted bright red. This was not for decoration, but so that... "a new hand, unused to scenes of strife, might not grow faint at the sight of blood splashed nearby."

· COAST GUARD SERVICE ·

The United States Coast Guard was formed in 1915 by a combination of the Revenue Cutter Service and the Lifesaving Service. The motto of the Coast Guard Service is *Semper Paratus,* "Always Ready."

· ORGANIZER OF PILOTAGE · SYSTEM

George William Blunt, American hydrographer, made original surveys of many U. S. harbors. He helped promote the system of pilotage for New York City, and was influential in causing the Federal Government to inaugurate the Lighthouse Service.

· NAVY DAY ·

Navy Day was inaugurated by the Navy League in 1922. The Navy League, a civilian organization whose main objective is the welfare of Uncle Sam's seamen, chose October 27th as their date for commemoration, because that was the birthday of former President Theodore Roosevelt, who strove so diligently in promoting this country's modern Navy.

The Navy League has branches in almost one thousand communities, and strives ceaselessly to further the interest of the American public in their fighting ships and the men who man them.

Each year on October 27th parades and speeches are sponsored in all these communities as a token of respect to our naval forces and admiration of their fighting prowess.

Through the tireless efforts of the Navy League will come to many a seamen, standing a lonely watch in the far reaches of the globe, a spark of comfort . . . the knowledge that he is not forgotten.

WINDAS

• THE ACCIDENT WHICH PRODUCED A BOOK •

In 1839, a stage coach accident was directly responsible for the publishing of "Maury's Sailing Directions." Badly hurt in the accident, Captain Matthew Fountaine Maury utilized the time he was compelled to stay in bed, by compiling his wonderful book on sailing directions and sea geography.

• SOLDIER ~ ING •

This term for loafing on a job is derived from very early times. Soldiers aboard ships would do their full share of fighting, but refused to have anything to do with the working of the vessel, feeling that it was beneath their dignity.

•FISH THAT COST A KING HIS HEAD •

Because Charles First put huge taxes on the great herring industry of England, it stirred up strife which finally resulted in Civil War...which in turn resulted in the beheading of Charles.

•PROVIDER ...on a BIG SCALE•

According to Irish law in the 13th. century, no commander of an Irish ship could put to sea without first having made sufficient provision for the wives and families of his seamen.

• SEA-GOING SNOW PLOWS •

In the far northern latitudes, plane carriers have found it necessary to add snow plows to the ship's gear.

The plows are used for keeping the flight decks clear of snow.

• LARGEST CONVOY •

Regarding the landing of American and British troops on the northern coasts of Africa during November, 1942, some idea of the vastness of the project may be gathered from the fact that over one hundred and twenty-five ships of war were required to shepherd the gigantic fleet of transports and supply ships through the danger zones of hostile seas.

The venture is considered the largest sea-borne invasion force of all time.

• THE INDESTRUCTIBLE • DESTROYER

A tiny destroyer with a crew of only seventy-three officers and men has become a tradition in Norway. She sank so much coastwise shipping that the Germans offered one million dollars reward for the capture of her captain, dead or alive.

This scourge of Nazi water-borne traffic, the *Sleipner*, generally attacked at night, roaring out of the darkness to sink or seize an enemy ship, then fading away into the gloom, leaving only a battered hull to mark her coming and going.

The *Sleipner* was bombed and shelled dozens of times. Again and again the Germans reported they had destroyed her, which speaks volumes for Nazi propaganda methods . . . for the *Sleipner* still sails the North Seas.

WINDAS

BUCCANEER'S ARTICLES

The first Workman's Compensation Insurance were articles signed by the buccaneers. These articles provided that in the event of wounds received in service, the victim should be compensated "For the loss of One Eye $500; for the loss of Right Leg $500"--- and so on through a list ranging approximately from $25 to $1,000.

FURL

The word "Furl"(as in, furl the sail) is derived from the old English "Fardle", meaning to make a bundle.

PEA-JACKET

This short coat or jacket was originally made of Pilot-cloth (material similar to Melton cloth) and was named for the initial letter of the word. Thus it was first spelled P-jacket, not Pea-jacket.

SALT-JUNK

Junk being a tough stringy fiber from which rope is made, one can readily understand and appreciate the nickname "Salt Junk," as applied to the beef supplied to the old Navy.

A LONG WAY TO THE BOTTOM

The greatest known depth in the Atlantic Ocean is 30,246 feet, at a point north of Puerto Rico.

• CARVEL-BUILT BOATS •

The building of boats by the carvel system is supposed to have been handed down from the Ancient Egyptians.

• FIRST PORT •

The first European structure in the Western Hemisphere was a fort built from the wreckage of Columbus' *Santa Maria,* which ran aground on the island of Haiti on Christmas Eve of 1492.

• OUR FIRST SEA-GOING • PRESIDENT

To Franklin D. Roosevelt goes the credit of being the first United States President to sail his own vessel to a foreign port.

This was when he took the schooner *Amberjack* to Campo Bello, Canada, on a vacation cruise which aroused world-wide interest.

To the seamen of our fleets it will present a common bond of mutual respect, for there is no spirit of free-masonry greater than that which binds together the brotherhood of the sea.

The Commander-in-Chief has demonstrated that not only is he deeply interested in matters maritime; he is qualified to be classed as a good sailorman, for he can "hand, reef, and steer."

· QUAKER GUNS ·

This was a name given to wooden guns in the 17th. century. Most ships carried a number of these dummy guns in addition to their regular batteries, in order to create an impression of being heavily armed, and thus discourage attacks by pirates.

·ARMY DIRECTED SEA-FIGHTS·

Until 1546, the supervision of a sea-fight was not entrusted to sea-faring men, but was under the direct control of some general or other army officer. Seamen were hired only to work the ship.

· QUARTER ~ DECK ·

The quarter-deck received its name in the days when decks were in tiers. The "Half-deck" was half the length of the ship, and the "Quarter-deck" was half the length of the half-deck.

· The 'INDIA RUBBER' MAN ·

Old Navy slang for the ship's physical instructor _ _ _ _ obviously because this trained athlete could twist, bend or stretch in any direction.

• IRISH HURRICANE •

Just old Navy slang for a dead calm.

• MODERN TYPE TORPEDO •

The first modern type torpedo was invented by Robert Whitehead in 1868.

• HAND GRENADES •

Hand grenades are supposed to be the invention of Chinese pirates. They consisted of short lengths of stout bamboo, packed tight with powder, and equipped with a wick similar to a modern fire cracker.

• TOUCHING THE STARS •

Here is a sentimental custom which seems to have started when Annapolis midshipmen were making their annual voyages to Europe.

The French and Scandinavian girls would touch the stars on the midshipman's dress uniforms, in the hope that it would bring them (the girls) good luck.

Nowadays the underlying thought has undergone a change. When an American woman (be she wife, mother, sister or sweetheart) is bidding farewell to her departing sailor, she quietly lays a finger on the stars ornamenting his uniform. This is done as a silent wish for his safe keeping and a happy return.

It is a simple custom, and like most things whose keynote is simplicity, it evidences faith and sincerity.

The passing of time will transform this modern custom into one of the tender traditions prized by all men who go down to the sea.

· JUNKS ·

These Oriental vessels got their name from English seamen's mispronunciation of the Chinese word "Cheung." On close inspection, the hulls of these little ships will be found to still adhere in almost every detail to the great Magellan's craft, from which they were copied. Even the sail plan is obviously a crude attempt to imitate the sails of the famous Portugese exploration ships.

· CREW ·

The members of a ship's company derive their name from the old Norse word "ACRUE." From this also comes our word "Recruit."

· MARTINET ·

This term for any strict disciplinarian was named for a French officer in Louis XIV's reign, who carried discipline to the nth. degree of harshness.

·1st. NAVIGATION BOOK·

The first book on navigation written in English was William Bourne's "Register of the Seas," published in 1573.

• FAMOUS PORT O' CALL •

Canton, China, has long been a favorite port with foreign seamen.

Its earliest notices date back to two centuries B.C. Arabs were sailing there as early as the ninth century, and the Portuguese by 1517. Then in the seventeenth century the Dutch came, and after them the English, and the Americans.

• HELL SHIP •

Times change, and so do the meanings of various terms, even among that most conservative and unchanging race of human beings, the deepwatermen.

For centuries the meaning of the word "Hell Ship" was very definitely that of a vessel whose master and bucko mates would drive and haze the crew to the limits of their endurance.

Volumes have been written around the subject, recording the almost unbelievable cruelties practised in the days of sail, when seamen were subjected to the merciless brutality of their officers.

But that was Yesterday.

Today the word "Hell Ship" refers exclusively to oil tankers bearing their highly inflammable cargoes through war zones. It is a terrible indictment against the piratical and savage ruthlessness of the German submarine campaign.

Hardly a tanker struck by a torpedo from these sea-going gangsters, but would spray burning gasoline and oil upon the seas in which honest seamen were struggling for their lives, as they abandoned their stricken ship.

How many of these hapless victims went to Davey Jones in a torment of flaming horror will never be known; but when peace comes at last, it might be well to remember why the meaning of the word "Hell Ship" was changed.

Especially when German merchantmen come seeking the sanctuary of United States and Allied ports.

WINDA5

· GALLEY WEST ·

"Knocked Galley West." meaning knocked cold
or stiff, was a term coined with reference to an old
Norse custom of placing the corpse of a Viking
chief in the sepulchral chamber aboard his galley,
and sailing her, afire, towards the setting sun.
The significance of the "cold" or "stiff" is of course
the corpse itself.

BELL~BOTTOM TROUSERS

Of all the reasons given for
the extreme width of sailors'
trousers at the bottoms, the
obvious and practical one
remains the best :- They were
easy to roll to the knees when
the owner was swabbing decks.

The FIRST COMPASSES

The first compasses known
were the crude though clever
contrivances which masters
of Arab dhows showed to
the sceptical Crusaders in
1190 A.D.

· FLAMBOROUGH ·

This English coastal town was
originally named "Flame-borough,"
because on the headland nearby
the city kept a beacon perpetually
burning to warn mariners away
from the dangerous shores.

• NORSE SHIPPING •

Norway's shipping industry is fifteen hundred years old.

• EAST INDIAMEN •

The big ships of the "Honorable John" (British East India) Company came into prominence early in the nineteenth century, although smaller ships of the same line had, of course, been trading to India years before that time.

But the later models were huge as compared to the earlier vessels, and were really a compromise between a man-o'-war and a merchantman.

They carried passengers and great cargoes. What is more to the point, they carried big crews trained to fight as navymen, and the ships were heavily armed.

This made them almost invincible to the pirates who infested the sea lanes that the East Indiamen used.

They were such a source of pride to the British public that young men of good families preferred service under the "Honorable John" house flag, to a commission in the Navy.

The masters of these ships had certain perquisites which masters of our modern liners might well envy.

As example: Passengers on the outward passage from England were required to furnish their own cabins: beds, chairs, tables, and everything else.

If these passengers did not make the return voyage, their goods and chattels became the property of the vessel's captain, who sold them at a very satisfactory profit to his next India-bound guests. Other "perquisites" went also to swelling the captain's purse, and instances are recorded where master mariners in this famous service retired in comfort after only two or three voyages. Anyway, they were great ships, and it is doubtful if any other line of sailing vessels ever left such an enduring reputation.

WINDAS

· DRINKING A TOAST ·

This term for drinking to one's health, or in one's honor, was coined in early days along the waterfronts, when it was customary to place a small piece of toast in the hot toddy and the mulled wine so popular with seamen of the day.

· AYE! AYE! ·

This seagoing affirmative was originally 'Yea! Yea!', but Cockney tongues twisted it to Yi! Yi! and eventually to it's present form, Aye! Aye!

· BUM-BOAT ·

A peddler's boat, filled with assorted merchandise. Its name is a corruption of "Boom-Boat," because these small craft were permitted to moor to a ship's boat-booms while disposing of their wares.

· KING-SPOKE ·

The spoke of a ship's wheel, which when perpendicular, indicates that the rudder is dead fore and aft, received its name from the old custom of decorating that spoke with a crown in honor of the king. . . . . .

• ANCHORS •

In the games of Euchre and Five Hundred, the jacks in the deck of cards are named for the main anchors aboard ship, the right and left bowers.

• LONG PIG •

This is a sinister term emanating from the cannibals of the South Sea.

Until the coming of the white man's ships, fish was about the only food other than fruit, roots, and berries, on the menu of the natives of the Coral Isles. With the white man came the pig, which increased in numbers to such an unbelievable extent, that among the islanders the word "pig" became a synonym for meat.

Therefore it was just a natural and logical thing to call a human victim prepared for cannibalistic consumption "long pig."

• LARBOARD SIDE •

This old-time name for the modern "port" or left-hand side of a vessel, means literally the "loading'" side.

Ancient ships were steered by a huge oar secured near the stern, on the starboard or right-hand side.

In order to keep the steering oar from being crushed against the side of the dock, ships were always tied up with their left-hand side fast to the wharf.

Therefore, as it was thus they were loaded, the left-hand side was called the "lar" or "load"-board side.

Later the name was changed to "port" side, because larboard was too similar to starboard, and created confusion when giving orders.

• GONE WEST •

This old term, referring to men killed in action, dates back to an ancient Viking custom of burial. When a Viking chief died, his body was placed in the sepulchral chamber aboard his ship. Then, with sail set, and steering oar lashed to keep her on her course, the vessel was set afire and pushed off into the west, toward the setting sun.

• PELORUS •

The navigation instrument for taking bearings is named for Hannibal's famous pilot, Pelorus, who made such a wonderful job of evacuating the noted Carthaginian general's troops from Italy after it became imperative to return to the defence of their own country......

• BLUNDER~BUSS •

A clumsy weapon, with trumpet-shaped muzzle, it received its name from the old Dutch naval "DUN-DER (Thunder) GUN." Notorious for inaccuracy, its name of "DUNDER" was changed to "BLUNDER." The "BUSS" is evidently a corruption of "BESS," after the English weapon of the period, "Brown Bess."

• **ANCHORS AWEIGH** • The word "Aweigh" is from the old English "Woeg" to raise.

• CARTEL-SHIP •

A cartel-ship is a vessel commissioned in wartime for the purpose of exchanging prisoners.

• THAT'S A "GALLEY" YARN •

Meaning in short, a doubtful piece of information from the cook's department, with the same ring of uncertainty that is implied in the more familiar phrase "that's a scuttlebutt rumor."

• FATHER OF MODERN • LIGHTHOUSES

To John Smeaton, a trained and highly skilled engineer who built the first really successful Eddystone Light in 1756, must be accorded the title "Father of Modern Lighthouses."

Like most master minds he encountered plenty of grief and opposition.

Masons refused at first to alter their old-time methods of cutting stone.

Skippers demurred at hauling the heavy granite blocks for the base of the lighthouse in their small vessels.

The people of Plymouth interfered with his labors for unknown reasons, and the press gang seized his workers to serve the King's ships.

But Smeaton refused to be dismayed, and, according to historians, the job was finished and candles were lighted in the lantern of the Eddystone in 1759.

The light guided mariners past the dangerous coast for one hundred and twenty years, and though it has been replaced by the present modern structure, the fame and perseverance of Smeaton still endures with undiminished brilliance.

WINDAS

· TAPS ·

The Last Call got its name from the Dutch word TAP-TOE, meaning to turn off all beer spigots (or taps) and to put out all lights in waterfront taverns. From this same word we get also the corrupted term "Tat-too."

·HE GETS the POINT·

Following a courtmartial a British officer knows if he has been found guilty by the position of his sword which is placed on a table before him. If the point faces him he is guilty; if the hilt is nearest, he is innocent.

·SWORD SALUTE·

The first movement of the modern sword salute is a survival of the ancient custom of kissing the cross which was emblazoned on the hilt of every Christian's sword.

· WOMEN SPONSORS

Until 1846, it was thought 'un-ladylike' for the fair sex to smash bottles of wine against ships' noses. However, in October of that year, Miss Watson of Philadelphia established a precedent by sponsoring the launching of the U.S.S. Germantown, and ever since then our fighting craft have been christened by the ladies.

• BUGLE •

This instrument received its name from medieval French sailormen who blew loud blasts on ox horns when celebrating a victory. The word "Bugle" in old French literally means "Wild Ox."

• DOG-WATCH •

The name for the split watch between the hours of four to six and six to eight p.m., was originally "Dodge Watch," as it allowed seamen to escape (or dodge) standing the same watch every day of the voyage. As time went on the name was gradually corrupted to the present "Dog-Watch."

• WARD-ROOM •

The ward-room originally was known as the WARD-ROBE ROOM, being the place where officers kept their spare wearing apparel and also any loot they won while on service. It was not until years later that it served its present purpose and became the officers' mess-room.

• MAROONED •

This old punishment for mutineers consisted of placing them on an island with musket, cutlass, and a breaker of water, and leaving them to their fate. It got its name from certain Ci-maroon Indians who had been transplanted to the West Indies as cheap labor and, deserted by their Spanish masters, had been left to starve to death. The famous Captain Drake discovered them in a pitiable condition and gained the Indians' lasting gratitude by returning them to their far-off home.

· FATHOM ·

This well-known nautical word comes from the old English "FAETM" meaning to embrace. Parliament decided that, since an embrace involved the distance between a man's hands when placed around his sweetheart, and as that distance averaged about six feet, it should be established as a standard measure.

· AHOY ·

This old traditional greeting for hailing other boats, was originally a Viking battle-cry.

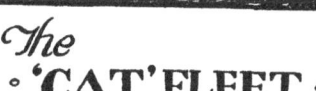

The · 'CAT' FLEET ·

A nickname for those British squadrons or fleets, led by a ship of the "LION" or "TIGER" class, during the World War.

· SHIP-SHAPE ·

Just in case you think this is a term of <u>recent</u> origin, please note that Xenophone coined this phrase while chatting with his boy friend Socrates.

And now, as all things must, this book comes to an end. If the reader has had just one half the enjoyment in the reading, as the writer had in the preparation and execution of the volume, no one will regret the time thus mutually expended.

To the men and the ships that contributed their valiant share to our traditions, we drink a deep-sea toast.

To those men and their ships yet to come, who will carry on and create new traditions, so that our sons and their sons may take great comfort in the Brotherhood of the Sea, we give the time-honored hail and farewell "Fair Winds and a Safe Voyage."

So fare ye well, my bonnie gal
I can no longer stay;
For I'm off on a trip in a government ship
Ten thousand miles away.

Old Sea Shanty

Finis